Receptors in the Evolution and Development of the Brain

Matter into Mind

Receptors in the Evolution and Development of the Brain

Matter into Mind

Richard E. Fine

Professor Emeritus, Department of Biochemistry,
Boston University School of Medicine,
Boston, MA, United States

ELSEVIER

ACADEMIC PRESS
An imprint of Elsevier

Academic Press is an imprint of Elsevier
125 London Wall, London EC2Y 5AS, United Kingdom
525 B Street, Suite 1650, San Diego, CA 92101, United States
50 Hampshire Street, 5th Floor, Cambridge, MA 02139, United States
The Boulevard, Langford Lane, Kidlington, Oxford OX5 1GB, United Kingdom

British Library Cataloguing-in-Publication Data
A catalogue record for this book is available from the British Library

Library of Congress Cataloging-in-Publication Data
A catalog record for this book is available from the Library of Congress

ISBN: 978-0-12-811012-6

For Information on all Academic Press publications
visit our website at https://www.elsevier.com/books-and-journals

Publisher: Nikki Levy
Acquisition Editor: Natalie Farra
Editorial Project Manager: Kristi Anderson
Production Project Manager: Sujatha Thirugnana Sambandam
Cover Designer: Mark Rogers

Typeset by MPS Limited, Chennai, India

Contents

Acknowledgments

I would first like to acknowledge the contribution of Joshua Rubin, MD, PhD, with whom I wrote the article that provided the outline for this book. I am thankful to Dr. Susan Leeman for introducing me to the wonders of neuroscience; and to my scientific mentors, Drs. Dennis Bray, Sidney Brenner, and Francis Crick. Dr. Maxwell Cowan first described apoptosis in the developing nervous system. He also offered to grant me a Howard Hughes Award if my dean nominated me. Dean Aaron Chobanian refused and provided the initial impetus to write this book. I am indebted to my collaborators during the last 40 years including Drs. Blitz, Rothman, Herschman, Rome, Rotundo, Connor, Kosik, Wells, and Morin, and to the 30 graduate students and postdoctoral fellows who taught me a great deal about respect and compassion. An untold number of scientists' work made this book possible. The cartoon used in this book was created by Bryce Howat. I am indebted to Dr. Alan Blitz for critically reading an early version of the book. I owe a great deal to my editorial project manager, Kristi Anderson, for her cheerful advice and support. I was helped immensely to muddle through by Dennis, Marni, and John. Rosalind put up with me for many years. I owe it all to the support of my sons, Mark and Eric. Finally, to my wife Lois, I acknowledge her support, encouragement, and most of all, her love during the writing of this book.

Introduction

This book has been written with three groups of readers in mind. The first group consists of undergraduates, graduate students, and postdoctoral fellows in neurobiology and psychology. They can employ the book as a supplement together with a comprehensive neuroscience textbook such as Squire et al. The second group consists of first-year medical students who use this book to accompany a medical neuroscience textbook. Finally, this book may prove interesting to a lay audience desiring to understand how the human brain develops.

A number of figures have been used to provide more detailed information on topics discussed in the text. They are not essential to understanding the text, and can be left out without lessening an understanding of the topics described in the book.

Chapter 1, Classes of receptors, their signaling pathways, and their synthesis and transport, is a description of the key groups of receptors and ligands that play key roles in brain development, as well as their signaling pathways and modes of synthesis and transport. This chapter can be skipped by readers who already have an understanding of the cell biology of receptors.

Finally, a list of recommended textbook chapters and review articles are placed at the end of each chapter. They also represent more detailed information and are not essential readings.

Classes of receptors, their signaling pathways, and their synthesis and transport

1

CHAPTER OUTLINE

The human brain is the most complex organ ever assembled. It is composed of at least 100 billion neurons and about the same number of glial cells, which communicate through a network of trillions of chemical synapses and electrical channels called gap junctions, as well as other cell types such as endothelial cells that are present in much smaller numbers. This conglomerate of connections can solve incredibly difficult mathematical problems, compose symphonies, communicate with other brains either by speech or writing, build rockets to reach the moon, hit a baseball moving at 100 mph, and produce incredibly beautiful paintings.

Perhaps the most amazing thing about the human brain is that it can build itself from a small group of undefined cells generated early in development. The purpose of this book is to outline a plausible strategy for the development of

Receptors in the Evolution and Development of the Brain. DOI: https://doi.org/10.1016/B978-0-12-811012-6.00001-7

the human brain. Central to this strategy is the use of two classes of molecules, receptors, and ligands—a scientific term for molecules to which receptors bind with great specificity and selectivity. Once a receptor binds to its ligand, it changes its shape, thereby sending a signal to the cell interior. This fundamental mechanism has evolved from the most ancient unicellular organisms, Archaea, and has been critical in the evolution of multicellular organisms and in the origin of the brain, leading to the development of the most complex assemblage of all, the human brain.

This book is entitled *The Key Role of Receptors in the Evolution and Development of the Brain: Matter into Mind.* To understand what this key role is, one must understand the unique properties of receptors. In this chapter the reader will be acquainted with these properties before moving into the central areas of the book, the evolution of receptors and organisms and the concomitant evolution of neurons leading to the most complex system that we know of: the mammalian brain. (Humans are thought to have the highest intelligence. However, one can argue that whales and elephants, which have brains containing more neurons and glia than humans, are also more intelligent. Perhaps their greater intellectual ability makes them unwilling to destroy our planet.)

Receptors are proteins found at the outer membrane of the cell, the plasma membrane, or within the cell either in the cytoplasm or the nucleus. An individual receptor has several key characteristics. It is able to recognize and tightly bind a very specific molecule including small molecules, another protein, a lipid, or a sugar. This selective tight binding is a key feature of the receptor—ligand interaction. A receptor can bind to a specific molecule when the latter's concentration is 10−12 molar or lower. At the same time it will not recognize and bind to a very similar molecule that is present at a million-fold higher concentration.

Upon binding to its specific ligand, the receptor changes its shape—the scientific word is *conformation.* This in turn causes information to be transferred to other molecules in the cytoplasm of the cell, leading to an almost infinite variety of changes in the behavior of the cell. These include alteration in the direction and/or the speed of movement, increases in the type and the amount of various proteins produced by the cell, and an alteration of the electrical excitability of a neuron. In the most extreme cases, receptor—ligand binding leads to the suicide of the cell, also called apoptosis. This topic will be discussed in detail in Chapter 9, The importance of rapid eye movement sleep and other forms of sleep in selecting the appropriate neuronal circuitry; programmed cell death/apoptosis.

While there was pharmacological and physiological evidence that specific receptors existed for quite some time, the first isolation of a receptor, the nicotinic acetylcholine receptor (NAchR), occurred in 1972. The NAchR from the eel electric organ, a modified muscle that contains many NAchRs, was used. The isolation was based on the coupling of a small molecule, α-bungarotoxin, which specifically bound to this receptor, to a resin poured into a column. Only the NAchR specifically bound to the immobilized α-bungarotoxin. After washing several times the receptor was eluted with a mild detergent, which preserved its structure. It was then incorporated into an artificial membrane and shown to have the pharmacological properties identical to the NAchR. This method has been used with modifications to purify many important receptors.

1.1 TYPES OF RECEPTORS

As expected for a class of molecules that carries out a myriad of functions and can bind to a huge number of divergent molecules, there are many types of receptors. In fact, receptors comprise the most abundant group of proteins in the human genome. However, they can be organized into a relatively small number of classes, suggesting that the members of each group evolved from a single receptor.

G PROTEIN LINKED RECEPTORS

The most numerous class of receptors is called the G protein coupled receptors (GPCRs), referring to the ability of these plasma membrane receptors to bind to a cytoplasmic protein called the G protein after binding to an extracellular ligand. The G protein subsequently transfers information into the cell. A large percentage of drugs are directed toward the GPCRs.

All of the GPCRs contain seven α-helical segments or protein domains composed of hydrophobic ("water hating") or neutral amino acids that traverse the plasma membrane and form a hydrophilic ("water loving") channel in the middle. The α-helix, whose structure was first solved by the Nobel Prize winning chemist Linus Pauling, is ideal to pass through the very hydrophobic environment of the plasma membrane. The α-helix, when composed of hydrophobic amino acids, can neutralize all of its charged peptide bonds by hydrogen bonding to another amino acid 3.6 amino acids away in the α-helix. Fig. 1.1A. It requires 20−21 either hydrophobic or neutral amino acids to completely traverse the 35-angstrom hydrophobic environment of the bilayer portion of the plasma membrane.

The amino acids comprising bacteriorhodopsin like those in the GPCRs, contain seven transmembrane α-helices that cross the plasma membrane (Fig. 1.2). Bacteriorhodopsin was the first membrane protein whose three-dimensional structure was determined.

Another protein domain whose structure allows it to traverse the plasma membrane is the β-pleated sheet (Fig. 1.1B−F). However this structure is primarily used by protists, which include bacteria and other single-celled organisms. β-pleated sheets also can form the cross β-pleated sheets, which are the insoluble structures called amyloid fibers.

As mentioned previously, the 35-angstrom portion of the plasma membrane as well as that of almost all cellular membranes is a bilayer composed of phospholipids interspersed with cholesterol molecules and glycolipids, which contain

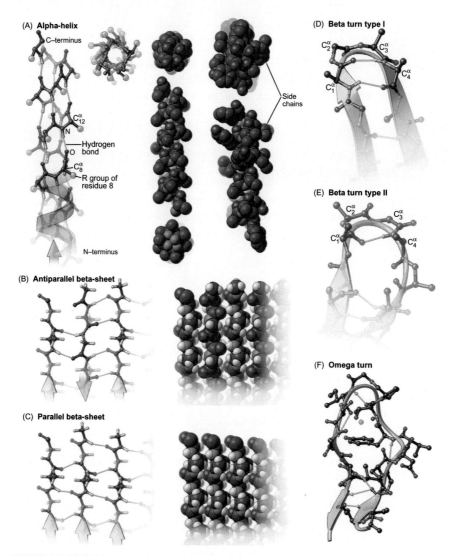

FIGURE 1.1 Models of secondary structures and turns of proteins.

(A) α-Helix. The stick figure (left) shows a right-handed α-helix with the N-terminus at the bottom and side chains R represented by the β-carbon. Hydrogen bonds between backbone atoms are indicated by blue lines. In this orientation, the carbonyl oxygens point upward, the amide protons point downward, and the R groups trail toward the N-terminus. Space filling models (middle) show a polyalanine α-helix. The end-on views show how the backbone atoms fill the center of the helix. A space filling model (right) of α-helix 5 from bacterial rhodopsin shows the side chains. Some key dimensions are 0.15 nm rise per residue, 0.55 nm per turn, and diameter of approximately 1.0 nm. (B) Stick figure and space filling models of an antiparallel β-sheet. The arrows indicate the polarity of each chain. With the polypeptide extended in this way, the amide protons and carbonyl oxygens lie in the plane of the sheet, where they make hydrogen bonds (*blue lines*) with the neighboring strands. The amino acid side chains alternate pointing upward and downward from the plane of the sheet. Some key dimensions are 0.35 nm rise per

(*Continued*)

complex sugars. The lipids that form the bilayer are also asymmetrically oriented with some phospholipids only on one side of the bilayer. Also some glycolipids form clusters on the extracellular side of the bilayer (Fig. 1.3). Integral membrane proteins including GPCRs completely traverse the bilayer; some proteins only extend through one layer of the bilayer. Others are present only on the membrane surface attached through electrostatic interaction (Fig. 1.3).

As expected for the most abundant class of receptors, members of the GPCR family carry out a huge number of different functions within the human nerve cells. The most abundant group is the odorant receptors, which can recognize more than 100,000 different odors. These cells, called nasal epithelia cells, are located in the nostrils. There are more than 1000 genes for odorant receptors in the human genome, comprising about 4% of the total genes.

Another very important group is the visual pigment GPCRs called rhodopsins. The three rhodopsins recognize different wavelengths of light in the photoreceptor cells of the retina.

Class C GPCRs, a third key class of GPCRs, recognize a whole host of different small peptides, including hormones and neuropeptides. Other members recognize small molecules that function as neurotransmitters, including glutamate (Glu), γ amino butyric acid (GABA), dopamine, and serotonin.

The β-adrenergic receptor was the first GPCR to be crystalized and its three-dimensional structure determined. Robert Lefkowitz won the Nobel Prize in Chemistry in 2012 for this achievement. Binding of the ligand causes a large change in the transmembrane portion of the receptor, thus activating the receptor. The structures of bacteriorhodopsin and the GPCRs are nearly identical.

In contrast to the GPCRs that bind large or small molecules and change their conformation resulting in cytoplasmic signaling, bacteriorhodopsin's function is to couple the light-induced passage of hydrogen ions from the cytoplasmic side of the membrane to the exterior. This produces the generation of a hydrogen ion gradient leading to the formation of the energy-providing molecule of the cell: adenosine triphosphate (ATP). ATP is the same energy-generating molecule employed by all organisms.

Bacteriorhodopsin is referred to as a receptor because it responds to a specific signal, light, by opening a channel through which H + ions can flow. Bacteriorhodopsin's seven α-helices comprise about 70% of the total number of

◀ residue in a β-strand and 0.45 nm separation between strands. (C) Stick figure and space filling models of a parallel β-sheet. All strands have the same orientation (arrows). The orientations of the hydrogen bonds are somewhat less favorable than in an antiparallel sheet. (D and E) Stick figures of two types of reverse turns found between strands of antiparallel β-sheets. (F) Stick figure of an omega loop.

Pollard TD, Earnshaw WC, Lippincott-Schwartz J, Johnson GT. Cell Biology. 3rd ed. Philadelphia, PA: Elsevier; 2016 [Chapters 1, 3, 13–16, 21–24, 27, 30, 33 and 34]. Fig. 3.8, p. 38. With permission.

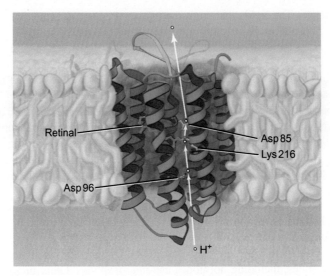

FIGURE 1.2 Proton pathway across the membrane through bacteriorhodopsin.

Numerous atomic structures, fast spectroscopic measurements of reaction intermediates, and analysis of a wide array of mutations revealed the pathway for protons through the middle of the bundle of seven α-helices. A cytoplasmic proton binds successively to Asp96, the Schiff base linking retinal to lysine 216 (Lys216), Asp85, and Glu204 before releasing outside the cell. Absorption of light by retinal drives conformational changes in the protein that favor the transfer of the proton across the membrane up its concentration gradient.

Pollard TD, Earnshaw WC, Lippincott-Schwartz J, Johnson GT. Cell Biology. 3rd ed. Philadelphia, PA: Elsevier; 2016 [Chapters 1, 3, 13–16, 21–24, 27, 30, 33, and 34]. Fig. 14.3, p. 38. With permission.

amino acids in the molecule as is true for many GPCRs. This roughly cylindrical structure forms an ion selective channel through the membrane (see Fig. 1.2).

While all members of the GPCR family have very similar structures to that of bacteriorhodopsin, possessing seven α helical transmembrane sequences, there is no amino acid sequence homology between them. It is likely that both the ancient Archaea, thought to be the first life on Earth, and the more recently evolved phyla including bacteria, plants, and animals have independently utilized this structure. A likely reason is its versatility, as discussed previously.

IONIC CHANNEL—FORMING RECEPTORS

A second major class of receptors that is very important in the brain is a quite heterogeneous group with ligand binding sites that, when occupied, open ion selective channels. These include the receptor for glutamate (Glu), the major excitatory neurotransmitter (Fig. 1.4). Other members of this family with similar structures include the γ amino butyric acid (GABA) receptor, which binds the

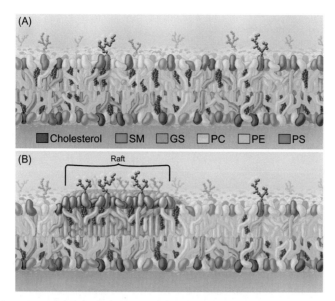

FIGURE 1.3 Asymmetrical distribution of lipids in the plasma membrane of an animal cell.

(A) Sphingomyelin (SM) and cholesterol form a small cluster in the external leaflet. *GS*, glycosphingolipid; *PC*, phosphatidylcholine; *PE*, phosphatidylethanolamine; *PS*, phosphatidylserine. PS is enriched in the inner leaflet. (B) Lipid raft in the outer leaflet of the plasma membrane enriched in cholesterol and sphingolipids. The lipids in the inner leaflet next to the raft are less well characterized.

Pollard TD, Earnshaw WC, Lippincott-Schwartz J, Johnson GT. Cell Biology. 3rd ed. Philadelphia, PA: Elsevier; 2016 [Chapters 1, 3, 13–16, 21–24, 27, 30, 33, and 34]. Fig. 3.8, p. 38. With permission.

major inhibitory receptor, GABA, and the nicotinic acetylcholine receptor (nAch), which binds Ach, the excitatory neurotransmitter at the neuromuscular junctions. These receptors, when they bind to their respective ligands, change conformations, allowing a channel to open. This in turn permits the passage of a charged ion, e.g., sodium (Na+) or calcium (Ca++), to enter the cytoplasm. The channel in these and in other receptors of this type is made up of α-helices from several different proteins. These receptors transmit information much faster than their GPCR counterparts, which usually work via their G proteins. (See section of this chapter on signaling).

Ionic Glu receptors consist of three major types: the AMPA receptor, the NMDA receptor, and the kainic acid receptor. The first two will be discussed here because of their key importance to brain development and function. Both the AMPA and NMDA receptor classes are composed of four similar but nonidentical subunits each having three membrane spanning α-helical segments (Fig. 1.4). AMPA receptors are for the most part permeable to Na+ after Glu binding, but a few also become permeable to Ca++ as well.

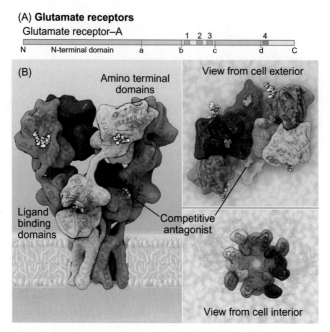

(A) Glutamate receptors

Glutamate receptor–A

N   N-terminal domain   a   b   c   d   C

(B) Amino terminal domains

View from cell exterior

Ligand binding domains

Competitive antagonist

View from cell interior

FIGURE 1.4 Glutamate-gated ion channel.

(A) Domain organization of the α-amino-3-hydroxy-5-methylisoxazole-4-propionate (AMPA)–type glutamate receptor with the glutamate-binding domain between a and b and four predicted transmembrane segments, M1 to M4. M2 is a P loop oriented toward the cytoplasm. (B) Ribbon diagrams of the atomic structure of the inactive AMPA-A channel with each of the four identical subunit polypeptides a different color. Space-filling models show a competitive antagonist bound to the glutamate-binding site. The side view shows the three domains. Note that the cytoplasmic domains of the blue and green subunits are oriented vertically, but that the yellow and red subunits swap their amino terminal domains. Top views show that the cytoplasmic domains have twofold symmetry, a dimer of dimers, while the transmembrane domains have fourfold symmetry.

Pollard TD, Earnshaw WC, Lippincott-Schwartz J, Johnson GT. Cell Biology. 3rd ed. Philadelphia, PA: Elsevier; 2016 [Chapters 1, 3, 13–16, 21–24, 27, 30, 33, and 34]. Fig. 16.11, p. 273. With permission.

The NMDA receptors normally are permeable to only Na + ions since a magnesium (Mg + +) ion sits in the opening of the channel, blocking access of the larger Ca + + ion. However, when the postsynaptic region containing the NMDA receptors becomes depolarized sufficiently, the Mg + + falls off the channel due to a conformational change in the receptor, and Ca + + can now enter the cell. Since Ca + + is a very important second messenger, it can modify the neuron in many ways as will be discussed in the signaling segment of this chapter.

GABAA receptors, the fast-acting GABA receptors, are composed of five similar peptides: two αs, two βs, and one γ. Each subunit has four membrane spanning α helices. When this receptor binds GABA it becomes permeable to Cl- ions, which hyperpolarizes the neuron, inhibiting its firing. During development, however, a different Cl- is present in GABAergic terminals making the GABA receptor excitatory rather than inhibitory. Another important homologous receptor found in the brain is the glycine receptor, which is another inhibitory ionic receptor.

Other important ionic receptors are the voltage-sensitive receptors that only open their respective channels when the voltage of the plasma membrane in which they are embedded rises from a negative voltage to near zero. One member of this receptor class is the Na+ channel, which is responsible for carrying the electric current in the axon from the cell body to the synapse of a neuron. Another important voltage-sensitive receptor is the K+ channel, which repolarizes the plasma membrane after the passage of the electric current down the axon. A third very important voltage activated receptor is the voltage-sensitive Ca++ channel in the presynaptic active zone, which becomes permeable to Ca++ ions after the presynaptic membrane is depolarized, triggering neurotransmitter release.

TYROSINE KINASE RECEPTORS

A third class of receptors is the plasma membrane embedded tyrosine kinases. Receptors as well as other proteins that can add phosphate groups to proteins, lipids, or sugars are called kinases. Tyrosine kinases are usually found in the plasma membrane as monomers with only one transmembrane α helical segment. When they bind to their ligand, they form dimers. In turn the cytoplasmically located kinase on one subunit adds a phosphate group to one or more tyrosine residues on the other member of the dimer. Once phosphorylated, these proteins change their shape and begin a chain of events in the cytoplasm leading to such changes as faster growth, differentiation, etc.

GLUCOSE TRANSPORTERS

A class of receptor that functions to bind a small molecule and then to transport it across the plasma membrane is the glucose transporter. Members of this receptor group are composed of one peptide containing 12 α helical, membrane-spanning domains that form a channel that selectively allows glucose to enter a cell down its concentration gradient. Several glucose receptors are very important in the brain, which derives almost all of its energy from the degradation of glucose. The Glut 1 glucose transporter allows glucose to enter the brain from the blood via the endothelial cells. It is also present on the glial cells of the brain. Glut 3 is only present on neurons, which are extremely active and demand a great supply of glucose to meet their energy needs. Finally, Glut 4 is present mainly on nerve cells in the cerebellum and motor cortex. It is not clear why this receptor, which is found on insulin or exercise responsive cells, adipose, and skeletal muscle respectively, is also present in neurons in the brain. One explanation may be that there is an increased energy demand on these neurons that are highly active during exercise. Therefore they require an additional source of glucose conveyed by Glut 4.

HISTIDINE KINASE RECEPTORS

Another class of receptors that are important in bacterium and plants but not in vertebrates are receptors that upon binding to their ligands become methylated on their cytoplasmic side. One result of this methylation is that the flagella on a bacterium reverse direction, allowing them to seek food or to flee from predators.

SERINE/THREONINE KINASES

Another group of receptors that is very important in early developmental processes, including that of the central nervous system (CNS), is comprised of serine/threonine kinases. These receptors also cross the plasma membrane only once. Two members of this receptor class are transforming growth factor receptor (TGFR) β1 and TGFRβ2. Two TGFβ molecules bind to two of each TGFRβ to form a hexametric complex. This complex phosphorylates serine and/or threonine residues on a group of proteins called SMADS. The TGFRs are very important in brain development as is the bone morphogenic protein receptor (BMPR).

TNF RECEPTOR FAMILY

Another important receptor class is comprised of transmembrane proteins that change conformation upon binding their ligand. This change induces the binding of information carrying proteins to their cytoplasmic surface. Among this group of receptors is the tumor necrosis factor receptor. Another receptor in this family that is very crucial in brain development is the low affinity nerve growth factor receptor, which will be discussed in Chapter 10, The key roles of BDNF and endocannabinoids at various stages of brain development including neuronal commitment, migration, and synaptogenesis.

STEROID HORMONE RECEPTORS

Besides the receptors anchored to the plasma membrane, a group of receptors including those for retinoic acid and the sex hormones, testosterone and estrogen, are nuclear molecules. Their respective ligands, hydrophobic small molecules that can cross the plasma membrane, then bind to them. This binding allows them to be activated and dimerize. The dimers bind to specific regions of DNA and activate the transcription into messenger RNA (mRNA) of various genes. A significant number of these mRNAs are important for many functions during neural development as will be discussed in detail in Chapter 13, Steroid hormones and their receptors are key to sexual differentiation of the brain while glucocorticoids and their receptors are key components of brain development and stress tolerance; the hypothalamic hormone producing cells: importance in various functions during and after brain development.

The retinoic acid receptor plays a crucial role in early neural development. Retinoic acid binding to its receptor is critical for ending the proliferative phase of neuronal development and inducing differentiation. Retinoic acid's precursor is vitamin E, and a vitamin E deficiency produces damage to the developing nervous system. After being converted into trans-retinol it becomes a crucial component of the visual pigments in the photoreceptor cells of the retina. Therefore, vitamin E deficiency is a leading cause of childhood blindness in developing countries where fresh fruits and vegetables, the major source of vitamin E, are unavailable.

NUTRITIONAL RECEPTORS

The previously mentioned classes of receptors are signaling receptors, conveying information from the outside to the inside of the cell. Another heterogeneous group of receptors has no signaling function. Instead they bind their ligands on the outside of the cell membrane and carry them into the interior of the cell, by a process called endocytosis, which is discussed below. In an acidic organelle called the endosome, the ligand and receptor separate. The receptor returns to the plasma membrane to be reutilized. In contrast, the ligand moves to a very acidic organelle called the lysosome. The lysosome contains a group of hydrolases that digest all the large macromolecules found in the cell. Any enzyme with an "-ase" ending belongs to a specific class of digestive enzyme, e.g., lipase, which degrades lipids, and protease, which degrades proteins. The resulting amino acids, lipids, sugars, nucleic acids, etc. serve as nutritional sources for the cell.

One nutritional receptor is the low-density lipoprotein receptor (LDLR), which brings cholesterol and triglycerides together with its ligand, Apolipoprotein B (ApoB), into the cell. This process of internalization is called clathrin-mediated endocytosis (Fig. 1.5). In response the cell stops synthesizing cholesterol. The brain, however, does not contain ApoB. Instead it employs another lipoprotein, Apolipoprotein E (ApoE), to carry cholesterol and triglycerides. A number of receptors related to LDLR bring these molecules into neural cells. The brain contains more cholesterol and other lipids than any other organ, so it is crucial for neural cells to possess this pathway. However, people who have the ApoE4 form are at much greater risk of getting Alzheimer's disease than people with the other two types of ApoE, that is, ApoE2 and ApoE3. This topic will be discussed in greater detail in Chapter 18, Trophic factor—receptor interactions that mediate neuronal survival can in some cases last through life, and trophic deficits can produce connectional neurodegenerative diseases.

Another important nonsignaling receptor is the transferrin receptor. Transferrin is a very abundant large protein found in the blood. It tightly binds all of the free iron in the blood, then binds to the transferrin receptor on the plasma membrane and is internalized into the cell by clathrin-coated vesicles (Fig. 1.6).

FIGURE 1.5 The intracellular processing and regulation of cholesterol biosynthesis.

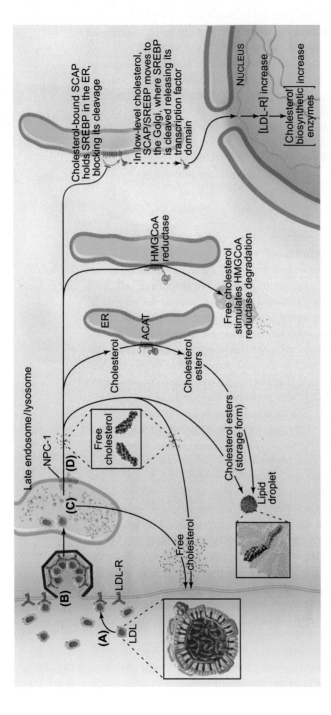

(A) Dietary cholesterol is delivered to cells in low-density lipoprotein (LDL) particles. (B) LDL particles are taken up by clathrin-mediated endocytosis. (C) Free cholesterol is released in late endosomes/lysosomes and transported to the cell surface or internal membranes, depending, in part, on the activity of the Niemann—Pick disease type C1 (NPC-1) integral membrane protein (D). Excess cholesterol can be acylated by acyl coenzyme A cholesterol acyltransferase (ACAT) activity and stored in cytoplasmic lipid droplets as cholesterol esters. ACAT activity is increased by high intracellular cholesterol levels. At the same time, high cholesterol in the membrane decreases new cholesterol synthesis by triggering the proteasome-dependent degradation of the enzyme β-hydroxy-β-methylglutaryl-coenzyme A (HMG-CoA) reductase. Finally, high cellular cholesterol decreases the uptake of LDL particles and dietary cholesterol by blocking proteolytic processing of the transcription factor sterol regulatory element-binding protein (SREBP), which is required for LDL-receptor (LDL-R) expression. Genetic defects that perturb steps A to D, which are required to maintain the delicate balance of cholesterol homeostasis, cause several human diseases. Familial hypercholesterolemia is caused either by a lack of LDL-R (A) or by LDL-R that is defective in endocytic activity (B). Wolman disease is a lysosomal storage disease that is caused by defective lysosomal cholesterol esterase activity; Niemann—Pick disease type C, another lysosomal storage disease, results in defective trafficking of cholesterol out of late endosomes and lysosomes caused by mutations in NPC-1 (D). *ER*, endoplasmic reticulum; *SCAP*, SREBP cleavage activating protein.

Pollard TD, Earnshaw WC, Lippincott-Schwartz J, Johnson GT. Cell Biology. 3rd ed. Philadelphia, PA: Elsevier; 2016 [Chapters 1, 3, 13–16, 21–24, 27, 30, 33, and 34].
Fig. 23.10, p. 403. With permission.

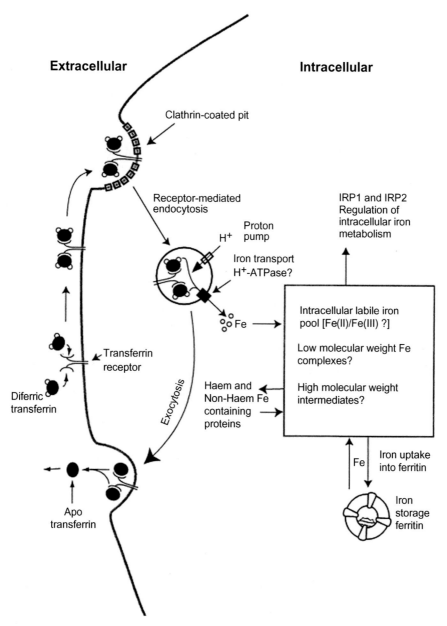

FIGURE 1.6 Schematic representation of Fe uptake from transferrin via receptor-mediated endocytosis in mammalian cells.

Extracellular diferric transferrin (Tf) is bound by the membrane-bound Tf receptor and internalized via receptor-mediated endocytosis into an endosome. Iron is released from Tf probably by a decrease in pH and is then transported through the membrane by an uncharacterized transporter. One candidate for this latter transporter in reticulocytes is the H + -ATPase, although the studies reporting this remain preliminary and require further

(Continued)

In the endosome, the iron is released from transferrin due to increased acidity. However, in contrast to the situation with ApoB, transferrin and its receptor stay bound to each other and are then recycled to the plasma membrane and the transferrin is released from its receptor to bind another iron molecule.

Since the brain cannot synthesize its own iron, it is dependent on a continuous supply of iron for a variety of purposes including the synthesis of ATP in mitochondria. Therefore, the brain endothelial cells that form the blood–brain barrier contain an abundance of transferrin receptors on the surface that faces the blood; these receptors endocytose the transferrin–iron complex and release the iron on the brain side of the barrier. It can then be endocytosed by neural cells. This topic will be discussed in greater detail in Chapter 6, Glial cells; the evolution of the myelinated axon; blood–brain barrier. Other plasma membrane receptors that are very important in the development of the brain are a large group of heterogeneous receptors that bind to large proteins in the extracellular matrix such as collagen or fibronectin. The receptors for fibronectin are called integrins and they are signaling receptors as well as extracellular matrix binding (ECM) proteins. Another important group of ECM binding receptors are the laminins, which can bind to integrins as well as to other receptors. Laminins are a key component of the developing brain ECM.

A major class of receptors that bind to a large molecule on the surface of an adjacent cell are the adhesion proteins. One important group is the immunoglobulin (Ig-CAM) family, which consists of hundreds of different molecules, many of which are expressed in the developing nervous system. These proteins all have repeating extracellular domains similar to those of immunoglobulins. They all bind to other ECM-associated molecules and some of their cytosolic domains bind to either actin filament–associated molecules or to signaling molecules.

Another very important group of adhesion receptors that play a major role in the development of the nervous system are the cadherins. Each cadherin has a single transmembrane sequence and they all contain several copies of the extracellular CAD domain. Each member of the cadherin family selectively binds to an identical molecule on an adjacent cell in the presence of a $Ca++$ ion. The cytosolic portion of each cadherin binds to two molecules, α- and β-catenin, which in turn are linked to the actin containing cytoskeleton. β-catenin, working through a different receptor, also plays a major role in an important intracellular signaling pathway, the Wnt pathway, which is discussed below.

◀ detailed investigation. Once the Fe has passed through the membrane it then enters a very poorly characterized compartment known as the intracellular labile Fe pool. ApoTf remains bound to the TfR and is then released by exocytosis. Iron that enters the cell can be used for metabolic functioning or can be stored in ferritin. It is thought that Fe in the intracellular Fe pool modulates the activity of iron regulatory proteins 1 and 2 (IRP1 and IRP2).

Richardson DR, Ponka P. Biochim Biophys Acta. 1997;1331:1–40. PMID: 9325434. Fig. 1. With permission.

Cadherins are key components in the early morphological fate decisions of the embryo. For example, as soon as the embryo forms three layers, the outer layer—called the ectoderm, from which the nervous system develops—expresses E-cadherin. Subsequently when the ectoderm forms the neural tube, the cells express N-cadherin.

A group with structural similarity to the cadherins is the protocadherins. The molecular diversity of this 80-member family results in many different combinations of tetramers that only recognize the identical tetramer on another cell. Therefore the protocadherins can specify over 3×1010 discrete combinations. These proteins, functioning as adhesion molecules, play a major role in the final specification of synaptic connections, which will be discussed further in Chapter 12, Key receptors involved in laminar and terminal specification and synapse construction.

1.2 RECEPTORS CONSIST OF A NUMBER OF FUNCTIONAL, STRUCTURAL, AND TOPOLOGICAL DOMAINS

Tyrosine kinases are an excellent class of receptors to illustrate the functional domain. The region of the tyrosine kinase that contains the kinase site is a functional domain. A structural domain is a contiguous sequence of 40 amino acids or more that assumes a distinct three-dimensional structure. The α-helix discussed previously is an example of a structural domain as is the β-pleated sheet. (See Fig. 1.1). These domains are used in many proteins, including receptors. An extremely interesting structural domain assumed by proteins under certain conditions is the crossed β-pleated sheet. This structure forms an extremely insoluble aggregate called amyloid fibers, which are likely to be involved in the pathogenesis of a number of connectional diseases of the brain; this will be discussed in Chapter 18, Trophic factor—receptor interactions that mediate neuronal survival can in some cases last through life, and trophic deficits can produce connectional neurodegenerative diseases.

The first two proteins whose structures were determined by X-ray crystallography, myoglobin and hemoglobin, have very similar structural domains and many α-helixes and β-pleated sheets are present in the similar structural region of each protein. The two proteins are therefore considered homologs and are very likely to have evolved from a common ancestor. The scientists John Kendrew and Max Perutz received the Nobel Prize in Chemistry in 1962 for their achievement in determining these three-dimensional structures. The GPCRs discussed previously are all homologs, as are many of the other classes of receptors described in the previous section.

Since the α-helix and the β-pleated sheet were described, nearly 100 other structural domains have been discovered by a combination of techniques including X-ray crystallography, nuclear magnetic resonance, and DNA sequencing.

Many of these structural domains, including the IgG and CAD domains mentioned previously, are found in neural receptors.

Structural domains in many cases have been preserved intact from primitive eukaryotes, or even Archaea. The amino acid sequence of a particular domain may change considerably but its structure remains very similar to the ancestral domain. These domains can be thought of like beads in a necklace consisting of many different beads. A given domain can be found in a number of proteins with different functions. In fact almost all the proteins coded for in the human genome consist of multiple structural domains, some of which are repeated many times. Fig. 1.7 shows a number of structural domains used in the important family of plasma membrane tyrosine kinases.

Finally, topological domains describe segments of proteins found on different sides of a membrane. A good example of this is the three components comprising plasma membrane-associated tyrosine kinases. They contain a ligand binding domain on the exterior side of the plasma membrane, a transmembrane domain, and a cytoplasmic domain containing the kinase structural and functional domains.

1.3 SIGNALING PATHWAYS FOR DIFFERENT CLASSES OF RECEPTORS

After a signaling receptor and ligand interact at the cell surface or, in the case of steroid hormone receptors, in the nucleus, the receptor changes its conformation. This in turn signals proteins in the cytoplasm of the cell to alter their functional states by moving, dividing, synthesizing new proteins, committing suicide, etc. Some of the signaling pathways that are important in the developing brain will be discussed in the following.

By far the most abundant and important group of receptors in the brain is the GPCRs described previously. The genes coding for the GPCRs comprise by far the largest group of genes in the human genome, over 1000 of the approximately 22,000 genes. As described earlier, the GPCRs are all structurally related to the Archaea protein, bacteriorhodopsin, in containing seven membrane-spanning hydrophobic α helical sequences surrounding a hydrophilic channel.

In the case of many of these receptors, upon binding to its specific ligand, the receptor changes its conformation and may form dimers. This conformational change causes a cytoplasmic trimeric G protein that consists of a Gα subunit combined with a Gβ, Gγ dimer to bind to its cytoplasmic domain. Following binding the GDP on the Gα subunit is released, causing the trimer to separate into an α subunit and a βγ subunit and leave the receptor. In the case of the β-adrenergic receptor, the separated α subunit now binds a GTP molecule. Then it binds to an inactive effecter, usually located at the plasma membrane, activating it. The Gα then hydrolyzes the GTP to GDP and rebinds to the βγ subunit. Fig. 1.8.

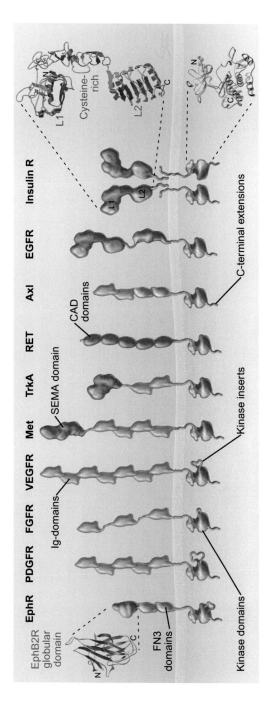

FIGURE 1.7 Receptor tyrosine kinases.

Domain architecture of 9 of the 20 families of receptor (R) tyrosine kinases, with ribbon models of several domains. The globular domain of the EphB2 receptor is a β-sandwich with a ligand-binding site that includes the exposed loop on the front of this model. The extracellular part of the insulin-like growth factor consists of two similar β-helical domains connected by cysteine-rich domains. The cytoplasmic kinase domain from the insulin receptor is similar to most typical kinases. Kinase inserts and C-terminal extensions contain tyrosine phosphorylation sites. Receptor names: *Axl*, receptor for the growth factor Gas6; *EGFR*, epidermal growth factor receptor; *EphR*, receptor for ephrin membrane-bound ligands in the nervous system, the largest class of receptor tyrosine kinases; *FGFR*, fibroblast growth factor receptor; *Met*, receptor for hepatocyte growth factor; *PDGFR*, platelet-derived growth factor receptor; *RET*, a cadherin adhesion receptor; *TrkA*, receptor for nerve growth factor; *VEGFR*, vascular endothelial growth factor. Domain names: *CAD*, cadherin; *F3*, bronectin-III; *Ig*, immunoglobulin.

Pollard TD, Earnshaw WC, Lippincott-Schwartz J, Johnson GT. Cell Biology. 3rd ed. Philadelphia, PA: Elsevier; 2016 [Chapters 1, 3, 13–16, 21–24, 27, 30, 33, and 34].
Fig. 24.4, p. 415. With permission.

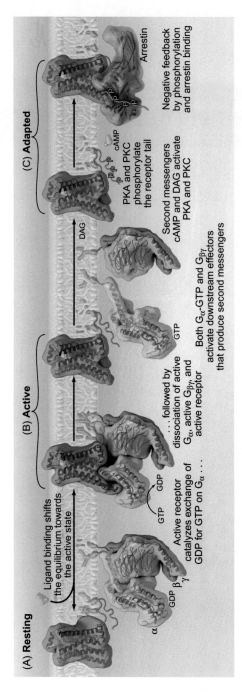

FIGURE 1.8 Activation and adaptation of a seven-helix receptor.

(A) Ligand binding shifts the equilibrium from the resting conformation toward the active conformation. (B) The active receptor binds trimeric G protein, and promotes dissociation of guanosine diphosphate (GDP), allowing GTP to bind. This dissociates Gα from Gβγ, allowing both to activate downstream effectors that produce, for example, the second messengers cyclic adenosine monophosphate (cAMP), which activates protein kinase A (PKA), and diacylglycerol (DAG), which activates protein kinase C (PKC). (C) Both kinases phosphorylate active receptors on their C-terminus, which attracts arrestin, putting the receptor into the inactive adapted state.

Pollard TD, Earnshaw WC, Lippincott-Schwartz J, Johnson GT. Cell Biology. 3rd ed. Philadelphia, PA: Elsevier; 2016 [Chapters 1, 3, 13–16, 21–24, 27, 30, 33, and 34]. Fig. 24.3, p. 415. With permission.

In the case of the serotonin receptor, the effecter is an enzyme, adenylate cyclase, which converts ATP to cyclic AMP (cAMP), a very commonly used second messenger. cAMP can in turn bind to a cytosolic enzyme called cAMP kinase that can phosphorylate many different proteins causing a variety of functional alterations in cell behavior. Second messengers also produce signal amplification. Thus the binding of one or a few ligand molecules can trigger a large cellular response. An important response triggered by cAMP kinase (PKA) is the phosphorylation of the cAMP response element-binding protein (CREB protein). This protein when phosphorylated migrates from the cytoplasm to the nucleus, where it activates the production of a number of new proteins. As will be discussed in Chapter 10, The key roles of BDNF and endocannabinoids at various stages of brain development including neuronal commitment, migration, and synaptogenesis.

Another second messenger system used by some GPCRs is composed of inositol trisphosphate (IP3) and diacyl glyceride (DAG), which are produced by the $G\alpha o$ or Gq containing G protein activation of phospholipase C. Phospholipase C catalyzes the hydrolysis of phosphoinositol diphosphate, producing IP3 and DAG. IP3 binds to the IP3 receptor on the surface of the endoplasmic reticulum (ER), opening a $Ca++$ selective channel allowing a massive release of $Ca++$ into the cytoplasm. The $Ca++$ in turn activates a number of $Ca++$ dependent kinases, some of which are activated by the $Ca++$ binding protein, calmodulin. The cytoplasmic $Ca++$ is rapidly pumped back into the ER by the action of a $Ca++$ ATPase located on the ER membrane.

DAG is the other second messenger molecule produced by the $G\alpha o$ or Gq catalyzed cleavage of phosphoinositol diphosphate. In combination with the $Ca++$ released from the ER, DAG binds to and activates an important enzyme called phosphokinase C at the plasma membrane. This enzyme in turn phosphorylates a variety of proteins that carry out many key cellular functions in the developing brain, including the activation of a number of transcription factors important for neural development. PKC also is involved in the biosynthetic pathways for prostaglandins and cannabinoids. The latter will be discussed at length in Chapter 19, Potential use of neuronal stem cells to replace dying neurons depends on trophic factors and receptors.

Another G protein involved in the signaling function of an important group of GPCRs is $G\alpha i$, $\beta\gamma$. This $G\alpha i$ subunit is used by an acetylcholine receptor, the α-adrenergic receptor, and by various neurotransmitter and taste bud receptors. Among the actions of $G\alpha i\beta\gamma$ that depend on the GPCR to which it is linked are inhibition of adenyl cyclase, opening $K+$ channels, and closing $Ca++$ channels.

Many GPCRs can be also inactivated by phosphorylation by protein kinases, which are activated by $G\alpha$ binding. This type of inhibition is referred to as "feedback inhibition." Also the phosphorylated GPCRs can bind a cytosolic protein called arrestin that triggers internalization of the receptor through clathrin-coated vesicle (CCV) formation. The internalized receptor can remain in the cytoplasm for a significant time before being returned to the plasma membrane or degraded. This process leads to long-term downregulation of the receptor.

Another important group of receptors that are activated by G protein subunits are the metabotropic GABA and Glu receptors. These receptors work over a longer time frame than do the ionic receptors, and they modulate the responses of neurons to these key fast-acting excitatory and inhibitory receptors.

1.4 TYROSINE KINASE RECEPTORS

As mentioned previously, a key group of plasma membrane embedded receptors are the tyrosine kinase receptors. These receptors have one membrane-spanning, ligand-binding extracellular sequence and their kinase site is located on the cytoplasmic side of the plasma membrane. When a ligand binds to its receptor, the receptor binds to another receptor molecule. Once the receptor dimer is formed, the two subunits phosphorylate each other. In the case of other tyrosine kinase receptors such the insulin receptor, the two subunits are already dimerized before ligand binding.

Once the subunits are phosphorylated they are recognized by various scaffolding proteins such as proteins containing either the SH2 or PTB domain (see below). Upon binding the SH2 domain, the effector molecule can propagate the signal through a variety of mechanisms. In one case the phospholipase C enzyme, which contains an SH2 residue, can hydrolyze phosphoinositide diphosphate, thereby producing two second messengers, IP3 and DAG. In another case the SH2 domain containing protein Grb2 combines with the protein SOS to activate the small GTPases, Ras and Raf GTPases, which causes another kinase, MEK, to phosphorylate MAP kinase, which enters the nucleus and phosphorylates a number of growth-promoting transcription factors. The neurotrophins including nerve growth factor (NGF) and brain-derived growth factor (BDNF) employ this signaling pathway. These proteins will be discussed in detail in Chapter 10, The key roles of BDNF and endocannabinoids at various stages of brain development including neuronal commitment, migration, and synaptogenesis.

1.5 SERINE/THREONINE KINASE RECEPTORS

As described previously, another group of plasma membrane receptors that are important in the development of the nervous system include the TGFβR and the BMPR. These receptors are serine/threonine kinases. Following ligand binding the subunits phosphorylate one another as well as a cytoplasmically localized protein called Smad. After phosphorylation the Smad protein enters the nucleus and activates a variety of genes involved in growth or differentiation.

1.6 SIGNALING THROUGH THE WNT PATHWAY

The Wnt signaling pathway is a critical pathway in the development of the brain. Wnt is a secreted protein that is modified by the addition of a fatty acid, making it more hydrophobic. Wnt binds to the receptor protein, frizzled, which is a member of the GPCR family. However this protein does not signal through a G protein. Instead it is associated with a coreceptor, LRP, a member of the low-density protein receptor family. When Wnt binds the receptor coreceptor complex LRP becomes phosphorylated and binds two proteins, axin and disheveled. Axin is thereby removed from a complex containing β-catenin, which when bound to axin and disheveled is targeted for degradation. When freed from the complex, β-catenin enters the nucleus and turns on the transcription of a number of mRNAs associated with cell division and differentiation. This pathway is called the canonical pathway. There are two other so-called noncanonical pathways that involve either release of Ca++ into the cytoplasm or translocation of the small GTPase Rho into the nucleus.

As mentioned previously, Wnt is hydrophobic because it has an attached fatty acid group, palmitic acid attached to its C-terminus. Therefore it can diffuse only a short distance from its point of secretion and has localized effects. As its concentration decreases it causes different effects on the cells to which it binds. Signals that induce differing effects in cells that receive large amount of Wnts from those that receive small amounts are called morphogens. These morphogens, which include Wnts, are critical in the early development of the brain and its segmentation into different regions as discussed in Chapter 8, The development of the cerebral cortex.

Another signaling system that is very important in the development of the brain is the Notch/Delta system. When the Delta ligand is not bound, the Notch receptor, a single membrane-spanning receptor, is adjacent to a protease, ADAM-10. Upon binding to Delta, which is also a single-pass plasma membrane protein on an adjacent cell, Notch is cleaved by Adam-10. The part of the receptor still inserted in the plasma membrane is cleaved within the plasma membrane by γ-secretase, the same enzyme that cleaves the C-terminal of the β-amyloid precursor protein (APP) to generate Amyloid β (Aβ) as well cleaving many other plasma membrane embedded proteins. APP and Aβ and their possible roles in Alzheimer's disease and Down syndrome will be discussed in Chapters 17, Role of trophic factors and receptors in developmental brain disorders, and 18, Trophic factor—receptor interactions that mediate neuronal survival can in some cases last through life, and trophic deficits can produce connectional neurodegenerative diseases. When Notch is cleaved, the cytosolic segment migrates into the nucleus, where it serves as a transcription factor for many mRNAs associated with cell growth and differentiation.

A third receptor ligand system that is critical to early brain development is the sonic hedgehog (Shh) pathway. Shh is cleaved from a precursor protein and a cholesterol added, making the molecule hydrophobic before it is secreted. Its migration is therefore severely limited, producing a concentration gradient. This behavior is characteristic of a morphogen that causes the morphogenesis of a discrete population of cells. Shh binds to a complex consisting of a ligand binding subunit, patched, and a signaling subunit, smoothened. Shh binding activates smoothened by blocking patch inhibition. This in turn activates a variety of signaling kinases and transcription factors, and the Gli transcriptional activator family of proteins.

1.7 SCAFFOLDING OR ADAPTOR PROTEINS HAVE ONE OR MORE CONSERVED MODULES

As discussed previously, some cytoplasmic or nuclear proteins recognize specific residues on other proteins such as the SH2 domain containing proteins that recognize phosphotyrosine residues. These proteins are members of a group of proteins called scaffolding or adaptor proteins. They themselves have no enzymatic or receptor activities. Instead they contain specific segments of amino acids that form very distinctive structural domains that have been conserved throughout evolution. Their function is to recognize specific structural domains on other proteins, bind to them, and then form signaling or structural complexes with other proteins including enzymes, transcription factors, etc.

Besides the previously mentioned SH2 domain and SH3 domains, other domains include the PDZ domains that recognize many different amino acid sequences on receptors and ion channel containing proteins. PDZ domain−containing molecules play key roles in organizing the pre- and postsynaptic endings and also in the specification of electrical junctions. These are discussed in Chapter 4, The growth cone: the key organelle in the steering of the neuronal processes; the synapse: role in neurotransmitter release.

Another important adapter molecule is the 14-3-3 molecule, which binds to phosphoserine residues in a number of important signaling proteins. As mentioned previously these include Ras and Raf, two small GTPases that are signaling proteins in the MAP kinase pathway used by receptor tyrosine kinases. The Trks, which are receptors for the neurotrophins, use this important pathway and will be discussed in Chapter 10, The key roles of BDNF and endocannabinoids at various stages of brain development including neuronal commitment, migration, and synaptogenesis.

Another family of adapter molecules that is important in the formation of CCVs comprises AP1, AP2, and AP3. These proteins serve as specific links between forming clathrin-coated pits and specific membrane proteins to be transported within them, following their budding to form CCVs. In the case of AP2, which is the best characterized and is involved in endocytosis, Fig. 1.9 demonstrates the complex cycle of CCV formation and uncoating. AP1 and AP3 mediate the formation of CCVs that bud from the *trans*-Golgi network (see below), and target the late endosome (AP1) or lysosome (AP3) respectively.

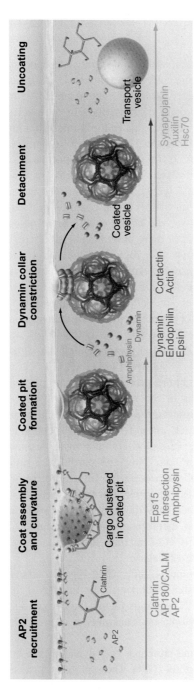

FIGURE 1.9 Timeline of receptor-mediated endocytosis driven by the clathrin-coated vesicle.

AP2 complexes are targeted to the plasma membrane and interact with tyrosine-based sorting motifs in the cytoplasmic domains of transmembrane proteins. AP2 also initiates the assembly of a polygonal lattice of clathrin, which concentrates cargo molecules in coated pits. AP2 and clathrin interact with the BAR-domain protein amphiphysin, which attracts the guanosine triphosphatase (GTPase) dynamin. After assembly of a coated pit and scission of the coated vesicle from the plasma membrane, auxilin and the adenosine triphosphatase (ATPase) Hsc70 dissociate clathrin and release the vesicle carrying cargo into the cell for fusion with endosomes.

Pollard TD, Earnshaw WC, Lippincott-Schwartz J, Johnson GT. Cell Biology. 3rd ed. Philadelphia, PA: Elsevier; 2016 [Chapters 1, 3, 13–16, 21–24, 27, 30, 33, and 34].
Fig. 22.9, p. 384. With permission.

1.8 SYNTHESIS AND TRANSPORT OF RECEPTORS AND TROPHIC FACTORS

Integral plasma membrane proteins as well as secreted neuropeptides and hormones follow a complex route before and after synthesis. As opposed to other proteins, these proteins are synthesized in a different manner. An mRNA for a particular membrane or secretory protein is transcribed in the nucleus and after maturation enters the cytoplasm. It contains an endoplasmic reticulum (ER) signaling sequence of nucleotides usually at the 5 prime end of the mRNA. Once the protein sequence begins to be translated from N to C terminal the 20−amino acid hydrophobic signal sequence is recognized by a cytoplasmic particle called the signal recognition particle. The signal recognition particle then attaches the partially synthesized protein together with the ribosome complex to the signal recognition particle receptor on the outer surface of the endoplasmic reticulum. While these steps are occurring, the addition of more amino acids is halted. The partially synthesized protein is then transferred to a region of the membrane called the translocon, which opens to form a small channel in the membrane in a process that requires the hydrolysis of GTP to GDP. The incomplete protein forms an α-helix as it moves to the luminal side of the membrane. Protein synthesis resumes, and in the case of a secretory protein, including neuropeptides and hormones, the whole protein enters the ER. While translation is still occurring, the signal sequence is cleaved by an ER luminal protein called the signal peptidase. A detailed description of these events is shown in Fig. 1.10.

Different types of integral membrane proteins including plasma membrane receptors employ different variations of the scheme shown in Fig. 1.10. In the case of type 1 integral membrane proteins, the N-terminal signal sequence moves through the membrane into the lumen as protein synthesis continues. At some point another 20−amino acid sequence called a stop transfer sequence is translated, which gets stuck in the membrane. Now the partially completed protein leaves the translocon and the rest of the protein is synthesized in the cytoplasm. The orientation of this common class of integral membrane protein has its N-terminal in the ER lumen, which becomes the extracellular space after vesicle transport and fusion with the plasma membrane, and its C-terminal in the cytoplasm (Fig. 1.11A). Members of this class include the human growth hormone receptor, the insulin receptor, and the LDLR.

Another class of integral membrane protein is the type 2 class. In this class there is a signal sequence/stop transfer sequence in the middle of the protein. It gets stuck in the membrane with its N-terminus facing the cytoplasm. The rest of the protein is synthesized in the cytoplasm. The transferrin receptor belongs to this class (Fig. 1.11D).

Another class of integral membrane protein is the glypiated proteins, which are anchored to a glycosylphosphatidylinositol (GPI). These proteins traverse the plasma membranes until a stop transfer sequence in encountered. A protease located in the

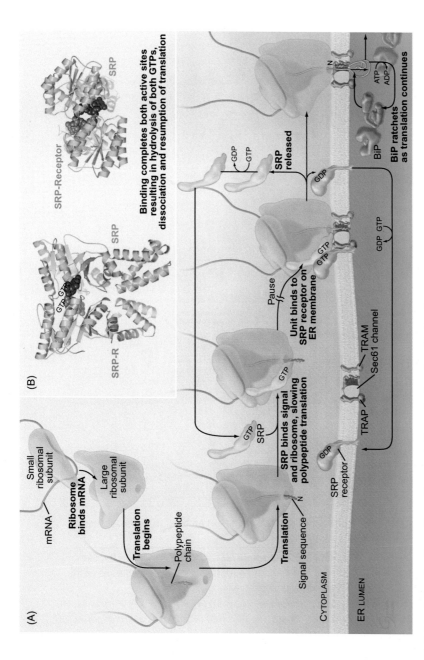

FIGURE 1.10 Cotranslational pathway from ribosome to the endoplasmic reticulum lumen.

(A) Signal recognition particle (SRP) and SRP-receptor use a cycle of recruitment and GTP hydrolysis to control delivery of a ribosome with an messenger RNA (mRNA) and nascent chain with a signal sequence to the Sec61 translocon in the ER membrane. SRP binds a signal sequence emerging from a ribosome and slows polypeptide translation. SRP also directs the ribosome to the SRP-receptor on the ER membrane, where the ribosome docks on the translocon and continues translation. (B) Ribbon diagrams of two views of the complex between SRP and SRP-receptor showing the close association between the two GTPase domains, which activates GTPase hydrolysis. *ADP*, adenosine diphosphate; *ATP*, adenosine triphosphate; *BiP*, binding immunoglobulin protein; *GDP*, guanosine diphosphate; *TRAM*, translocating chain-associating membrane protein; *TRAP*, translocon-associated protein.

Pollard TD, Earnshaw WC, Lippincott-Schwartz J, Johnson GT. Cell Biology. 3rd ed. Philadelphia, PA: Elsevier; 2016 [Chapters 1, 3, 13–16, 21–24, 27, 30, 33, and 34].

Fig. 20.6, p. 337. With permission.

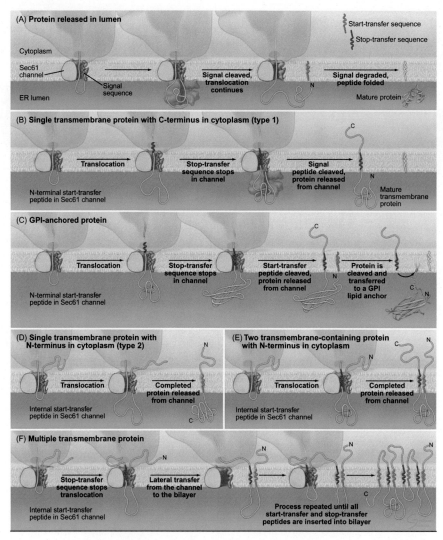

FIGURE 1.11 Comparison of pathways using SEC61 to target proteins to the endoplasmic reticulum lumen or membrane.

The drawings show how signal sequences, start-transfer sequences, and stop transfer sequences target protein to Sec61 channel and then to the lumen and membrane of the endoplasmic reticulum (ER). *GPI*, glycosylphosphatidylinositol.

Pollard TD, Earnshaw WC, Lippincott-Schwartz J, Johnson GT. Cell Biology. 3rd ed. Philadelphia, PA: Elsevier; 2016 [Chapters 1, 3, 13–16, 21–24, 27, 30, 33, and 34]. Fig. 20.8, p. 339. With permission.

ER membrane with its active site facing the ER lumen then cleaves the peptide and transfers the C terminal sequence to a GPI lipid (Fig. 1.11C). Isoforms of various tropic molecules including semaphorins and netrins belong to this class.

Another group of transmembrane proteins is similar to the type 2 class of proteins in that they both have signal/stop transfer sequences in the membrane. In contrast to the type 2 class the N-terminus as well as the C-terminus ends up in the cytoplasm (Fig. 1.11E).

A very important class of integral membrane proteins, to which the GPCRs belong, is the type 4 class. As described previously it has seven membrane-spanning domains. As it is synthesized the first hydrophobic sequence functions as a signal sequence, which brings the first group of synthesized amino acids into the ER lumen. Protein synthesis then occurs until a signal sequence is encountered, which brings the second group of amino acids into the lumen until a stop transfer sequence is encountered. This alternation between stop transfer and signal sequence alternates until seven transmembrane sequences are synthesized. The protein ends with its N-terminus in the lumen and its C-terminus in the cytoplasm (Fig. 1.11F).

Another class includes two proteins that play key roles in neurotransmitter release, as will be discussed in detail in Chapter 8, The development of the cerebral cortex. Vamp and Syntaxin as well as other proteins in this class are synthesized on cytosolic ribosomes with 20 hydrophobic amino acids at their C terminals. After synthesis they are transferred to a protein that binds to the C terminal peptides and escorts them to the ER. The hydrophobic peptides then pass through the membrane, stopping there while the rest of the proteins remain in the cytoplasm.

During their passage into the ER lumen, many integral membrane proteins and some secreted proteins have a string of carbohydrate groups consisting of two N- acetyl glucosamines, nine mannoses, and three glucoses added to one or more asparagine residues on the growing peptide chain. These groups get modified during passage through the secretory compartments; and after a brief labeling with a radioactive substance, provide a valuable tool to determine in which compartment a particular protein resides.

Even though all proteins possess the information in their amino acid sequences to form their appropriate three-dimensional structures, all but the smallest proteins will form nonspecific aggregates in the dense cytoplasm and ER luminal compartments. Therefore a group of protein chaperones have evolved that are evolutionarily very ancient, even found in the Archaea. These proteins use ATP hydrolysis to bind the unfolded polypeptides immediately after synthesis and hold them until they achieve their correct folding. Then they are released.

To prevent nonspecific protein folding and aggregation from occurring, the ER contains a large quantity of protein chaperones that bind to the immature proteins and only release them after they assume their correct conformations. Three of the most important of these molecules are binding immunoglobulin protein (BIP), calnexin, and calreticulin. BIP binds to the elongating peptides as they emerge into the ER lumen. It then releases the synthesized protein to either calnexin or calreticulin or both. These chaperones bind to the polypeptides until they form their correct conformations and then release them.

When a particular protein reaches its mature conformation, it is released. However, another problem that the newly formed peptide chains encounter as they exit the ER membrane and enter the lumen is that the lumen is an oxidizing environment as opposed to the cytoplasm, which is a reducing environment. This causes the cysteine residues in the peptides to form inappropriate intra- or inter-chain disulfide bonds. Two other important KDEL protein chaperones, protein disulfide isomerase (PDI) and ERp57, members of the thioreductase family, recognize these inappropriate disulfides and replace one cysteine residue breaking the inappropriate disulfide bond. The protein is then bound covalently to the enzyme until the correct disulfide forming cysteine containing peptide, either on the same protein molecule or on a different one, is found. The correctly folded protein is then released.

If the protein cannot assume its mature form due to a mutation or some other defect, the chaperone escorts the misfolded protein to a group of proteins collectively called EDAM, which send the misfolded protein back to the cytoplasm where it is digested in an organelle called the proteasome. The mechanism by which this retransfer across the plasma membrane occurs is not well understood.

Once the protein has the correct confirmation and disulfide bonds, it travels in a vesicle known as the COP II vesicle to the first compartment of the Golgi apparatus, a series of adjacent membranes directly opposed to the ER. in a vesicle known as the COP II vesicle. This Golgi compartment nearest the ER is called the *cis*-Golgi. Any of the KDEL chaperones that have escaped the ER are bound to the KDEL receptor and returned to the ER in COP I vesicles (Fig. 1.12).

After a protein reaches the *cis*-Golgi, some carbohydrates on the asn residues are modified. These proteins are called glycoproteins. The modifications continue as the glycoprotein proceeds sequentially through the medial and *trans*-Golgi stacks and finally to the *trans*-Golgi network.

In some cases another group of carbohydrates are added to either ser or thr residues. These modifications are called O linked and the proteins that are bound to these long chained carbohydrates are called proteoglycans.

The proteins are then trafficked sequentially to medial Golgi, *trans*-Golgi, and *trans*-Golgi network. Their carbohydrate groups are modified as they proceed. Once they reach the *trans*-Golgi network a major sorting step occurs, which is triggered by a slight drop in the pH. Proteins that are destined to be secreted or to become integral membrane proteins are clustered in particular regions of the *trans*-Golgi network and budded off in vesicles that are partially coated with clathrin or another as yet uncharacterized protein(s). The coat then falls off and the vesicles are transported via microtubules and kinesin motors (see Chapter 10: The key roles of BDNF and endocannabinoids at various stages of brain development including neuronal commitment, migration, and synaptogenesis) to the periphery of the cell, finally fusing with the plasma membrane.

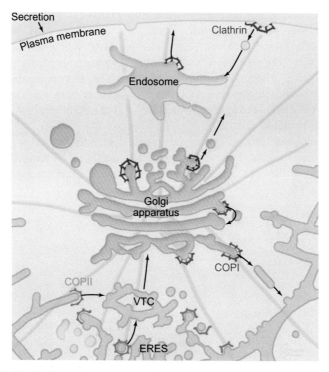

FIGURE 1.12 Distributions of coat proteins along the secretory pathway.

Clathrin, coat protein I (COPI), and coat protein II (COPII) coats are color coded.

Pollard TD, Earnshaw WC, Lippincott-Schwartz J, Johnson GT. Cell Biology. 3rd ed. Philadelphia, PA: Elsevier; 2016 [Chapters 1, 3, 13–16, 21–24, 27, 30, 33, and 34]. Fig. 21.7, p. 357. With permission.

    Other regions of the *trans*-Golgi network contain lysosomal enzyme precursors that are targeted by a signal sequence containing oligosaccharides with terminal mannose-6-phosphate groups to go to lysosomes. These proteins bind to the Mannose-6-Phosphate receptor, which recognizes these residues, and bud off the *trans*-Golgi network in CCVs. The small GTPase, Arf, is necessary for the budding to occur.

    Beside these two major destinations after leaving the *trans*-Golgi network, there are many other destinations to which vesicles are targeted. One can think of the *trans*-Golgi network as functioning like a cellular post office, directing each vesicle to its appropriate destination. In cells with regulated secretion, including neurons, the receptors and their associated ligands cluster at the ends of the *trans*-Golgi network and bud off forming regulated secretory vesicles. The regulated secretory vesicles require a signal, usually a large rise in cytoplasmic Ca++ for fusion with the plasma membrane or secretion to occur. Other material to be secreted or fused with the plasma membrane remains in the solute-rich lumen and buds off to form constitutive secretory vesicles.

A key family of small GTPases, the Rab family, is a critical element in maintaining this sorting network. There are more than 30 Rab proteins encoded in the human genome. Each one binds to the outer surface of its appropriate secretory vesicle. Upon reaching its proper destination it binds to a receptor(s) that is specific to the target membrane, triggering the hydrolysis of the GTP bound to the Rab to GDP. The Rab is then released from the membrane, the GDP is exchanged for GTP, and the Rab becomes ready to be used again (Fig. 1.13).

A very important process occurs in regulated secretory vesicles either before or after they are secreted. Secretory proteins all contain an N-terminal amino acid sequence called the pro sequence. This sequence is enzymatically cleaved

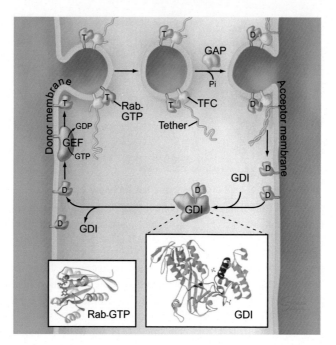

FIGURE 1.13 Rab GTPase cycle.

Rab-GDP is bound to GDI in the cytoplasm. A GTPase dissociation factor (not shown) on the membrane separates Rab-GDP from GDI so it can be activated by a membrane-associated, Rab-specific GEF. Rab-GTP recruits effectors, such as coiled-coil tethers and TFCs, which aid in targeting and docking the vesicle. A GAP stimulates GTP hydrolysis and returns Rab-GDP to GDI in the cytoplasm. Insets show ribbon diagrams of Rab-GTP and GDI. *GDI*, guanine nucleotide dissociation inhibitor; *GDP*, guanosine diphosphate; *GTP*, guanosine triphosphate; *GTPase*, guanosine triphosphatase; *Pi*, inorganic phosphate; *TFC*, tethering factor complex.

Pollard TD, Earnshaw WC, Lippincott-Schwartz J, Johnson GT. Cell Biology. 3rd ed. Philadelphia, PA: Elsevier; 2016 [Chapters 1, 3, 13–16, 21–24, 27, 30, 33, and 34]. Fig. 21.6, p. 357. With permission.

producing the final secretory peptide. Other important modifications that occur in the *trans*-Golgi network and/or the lumen of the secretory vesicles are the phosphorylation or sulfation of tyrosine residues.

The mechanism of vesicle fusion involves three integral membrane proteins. One is a member of the VAMP, otherwise known as the Synaptobrevin family, on the surface of the secretory vesicle. The second is a member of the Syntaxin family on the plasma membrane with which the vesicle fuses. The third is a member of the SNAP 25 family. All of these are specific to their appropriate secretory or plasma membrane. In the case of nonregulated secretion the two membranes fuse as soon as they touch. In the case of regulated secretion, fusion is prevented by a Ca++ activated protein, which is a member of the Synaptotagmin family. When the Ca++ concentration rises into the near-μmolar range, the target cell membrane and the vesicle membrane can now fuse (Fig. 1.14). The fusion of

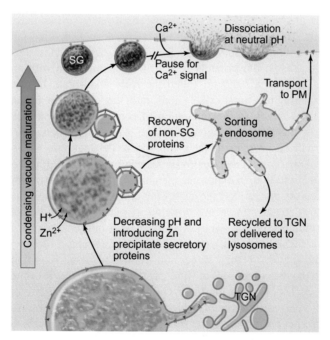

FIGURE 1.14 The vacuolar H + -ATPase in the secretory granule (SG) membrane lowers the internal pH.

This drives condensation and concentration of the contents. Dense-core, mature secretory granules are stored in the cytoplasm until a Ca2 + - mediated signaling event triggers fusion and release of their contents. Proteins inadvertently included in large, immature secretory granules emerging from the *trans*-Golgi network (TGN) are captured by clathrin-coated vesicles and recycled to endosomes and the TGN. *PM*, plasma membrane.

Pollard TD, Earnshaw WC, Lippincott-Schwartz J, Johnson GT. Cell Biology. 3rd ed. Philadelphia, PA: Elsevier; 2016 [Chapters 1, 3, 13–16, 21–24, 27, 30, 33, and 34]. Fig. 21. 27, p. 375. With permission.

synaptic vesicle, which is of great importance to the functioning of the brain, will be described in considerable detail in Chapter 9, The importance of rapid eye movement sleep and other forms of sleep in selecting the appropriate neuronal circuitry; programmed cell death/apoptosis.

1.9 METHODS OF RECEPTOR INACTIVATION

When integral membrane proteins including receptors carried in secretory vesicles reach the plasma membrane and fuse with it, the receptors have their luminal facing domain(s) exposed to the extracellular milieu. Receptors can now bind their specific ligands; and in the case of signaling receptors, activate a variety of signaling circuits. To prevent the signaling receptors from remaining activated for too long, one mechanism that is commonly employed is to internalize the receptor/ligand pair. They are then targeted to clathrin-coated pits that pinch off from the plasma membrane forming fully enclosed coated vesicles. (See Fig. 1.9). These vesicles almost immediately shed their coats and fuse with other internalized vesicles to form a structure called the early endosome. In the early endosome, an ATPase that pumps H + ions into the endosome is activated. The lowered (pH 6.0) leads to the separation of some ligands from their receptors. This structure functions analogously to the *trans*-Golgi network in targeting various internalized proteins to their appropriate destinations. As is true for the *trans*-Golgi network, the receptors to be recycled are clustered at the ends of the endosome and return to the plasma membrane. The soluble contents, together with any receptors that are not recycled, stay in the lumen.

The ligands in the lumen are targeted to multivesicular bodies and move into the perinuclear region. They then fuse with late endosomes and become even more acidified (pH 5.0) (Fig. 1.15). Some receptors are removed from the late endosomes at this stage and return to the *trans*-Golgi network contained in coats composed of five proteins called the retromer. One key protein that cycles between the *trans*-Golgi network and the late endosomes is the mannose-6-phosphate receptor, which binds to mannose-6-phosphate groups on lysosomal enzyme precursors in the *trans*-Golgi network (Fig. 1.15). A significant fraction of internalized transferrin receptors use this route as well to return to the plasma membrane. Finally the late endosomes become even more acidic (pH 4.0) to form or fuse with preexisting lysosomes, a major cellular digestive organelle (Fig. 1.15). Some receptors stay attached to their ligands after internalization and are also digested in the lysosome. One example is the epidermal growth factor receptor, which in some cell types is digested in lysosomes, and in others, recycles to the plasma membrane (Fig. 1.15).

Another important mechanism used by the cells to inactivate GPCRs is to reattach the G protein subunits that have become separated from the GPCR after

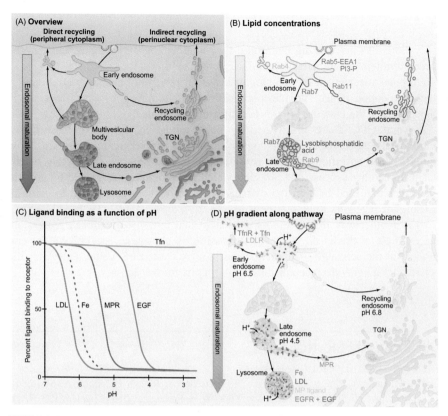

FIGURE 1.15 Membrane traffic and content sorting along the endocytic pathway.

(A) Overview. Cargo and membrane taken up by clathrin-mediated endocytosis are delivered to tubulovesicular early endosomes, which mature into multivesicular bodies and late endosome before fusion with lysosomes. Each compartment sorts membrane-containing receptors and other proteins into tubules and vesicles that are recycled to the plasma membrane either directly or indirectly through perinuclear recycling endosomes or the *trans*-Golgi network (TGN). Endosomes mature by accumulating internal membranes, delivery of lysosomal hydrolases from the TGN, and acquisition of targeting and fusion machinery. (B) Domain organization of the endocytic pathway. Local production of PIs in subregions of endosomal membranes recruit specific Rabs and Rab-binding proteins such as EEA1 (depicted in different colors). (C) Interactions of many cargo molecules with their transmembrane receptors depend on pH. (D) V-type rotary ATPase proton pumps progressively acidify the lumens of endosomal compartments. This gradient of pH facilitates protein sorting by regulating where along the pathway that each ligand dissociates from its transmembrane receptor. Dissociated ligands in the luminal space are targeted to lysosomes, whereas receptors in the membrane can be recycled to the cell surface in tubular endosomes. Iron dissociates from transferrin (Tfn) in early endosomes and apotransferrin without bound iron recycles with its receptor. Mannose-6-phosphate proteins dissociate from their receptors (MPRs) in late endosomes. Some ligands, such as epidermal growth factor (EGF), remain bound and are delivered with their receptor [epidermal growth factor receptor (EGFR)] to lysosomes.

Pollard TD, Earnshaw WC, Lippincott-Schwartz J, Johnson GT. Cell Biology. 3rd ed. Philadelphia, PA: Elsevier; 2016 [Chapters 1, 3, 13–16, 21–24, 27, 30, 33, and 34]. Fig. 22.12, p. 388. With permission.

activation, thus inactivating them until they bind another ligand. This requires the three G protein subunits to first reattach to one another.

Another mechanism of receptor inactivation involves phosphorylation and/or dephosphylation. Once the phosphotyrosines are removed by specific tyrosine phosphatases, the tyrosine kinase receptors are inactivated.

Several GPCRs are inactivated by phosphorylation. In one process called homologous desensitization, rhodopsin and the β-adrenergic receptor are phosphorylated on specific set of thr residues and the GPCRs are completely inactivated. In another process, the phosphorylated receptors are recognized by arrestin molecules that bind to both the cytoplasmic side of the receptor and to clathrin, causing the receptors to be internalized.

Yet another way that receptors are inactivated is termed feedback repression. An example of this is the feedback repression by phosphokinase (PKA), the end product of the activated β-adrenergic receptor. Activated PKA phosphorylates the activated β-adrenergic receptor thus inhibiting its further activity.

GENERAL REFERENCES

Recommended Book Chapters

Lodish H, Berk A, Kaiser CA, Krieger M, Bretscher A, Ploegh H, Amon A, Martin K. *Molecular Cell Biology*. 8th ed. New York: W.H. Freeman; 2016 [Chapters 1, 3, 7, 11, 13–18].

Pollard TD, Earnshaw WC, Lippincott-Schwartz J, Johnson GT. *Cell Biology*. 3rd ed. Philadelphia, PA: Elsevier; 2017 [Chapters 1, 3, 13–16, 21–24, 27, and 30].

Review Articles

Richardson DR, Ponka P. The molecular mechanisms of the metabolism and transport of iron in normal and neoplastic cells. *Biochim Biophys Acta.*. 1997;1331:1–40.

Wang L, Wang X, Wang CC. Protein disulfide-isomerase, a folding catalyst and a redox-regulated chaperone. *Free Radic Biol Med.*. 2015;83:305–313.

The RNA world: receptors and their cognate ligands in Archaea, Bacteria, and Choanoflagellates

2

CHAPTER OUTLINE

2.1 THE RNA WORLD

The Earth before the dawn of life was a hellish place. There was no oxygen in the atmosphere, periodic asteroids and comets struck the surface of the planet, and there were periodic huge volcanic eruptions. There was water, however, and this allowed the formation of all the necessary building blocks for life—including nucleotides, amino acids, fatty acids, and sugars. Experiments in test tubes have demonstrated that these molecules could be synthesized under conditions existing during the first 500 million years after the Earth's formation.

There are two components that are essential for producing life, that is, a living cell. One is a self-replicating molecule. The second is a membrane that separates the interior of the cell from the outside environment. Over the last 40 years, scientists have discovered that an RNA molecule is a strong candidate for the first replicating molecule. The most important evidence supporting the RNA first hypothesis is the work of Thomas Czech and Sidney Altman. They demonstrated that in a test tube, a specific type of RNA functions as an enzyme called a ribozyme. For this discovery they shared the Nobel Prize in Chemistry in 1989. This discovery demonstrates the possibility that RNA can carry out the steps necessary for its own replication. However, self-replicating RNA has not been found in a living organism. Nonetheless it remains the most plausible hypothesis.

The second component necessary for life is a suitable membrane that protects the contents of the cell from leaking out, but is permeable to the molecules

Receptors in the Evolution and Development of the Brain. DOI: https://doi.org/10.1016/B978-0-12-811012-6.00002-9

necessary for the cell to have a source of energy and to divide. The group of molecules that achieve these requirements are fatty acids composed of various length carbon chains. These fatty acids can form bilayer containing membrane vesicles similar to the plasma membranes found in living cells. The fatty acid bilayer is quite impermeable to most charged molecules. The resultant vesicles are also permeable to nucleotides and other components essential for replication. The fatty acid containing membranes can grow and also divide in aqueous solution at physiological pH and temperature (Fig. 2.1).

Amino acids were definitely present in the prebiotic world as well as some small peptides. It does appear, however, that the first functional peptides including some receptor-like molecules were synthesized by primitive ribosomes consisting only of RNA. In fact it is still true in all higher organisms that the

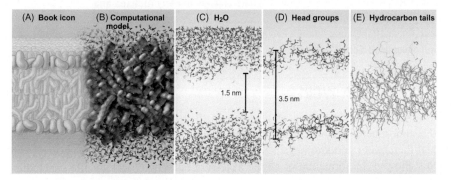

FIGURE 2.1 Computational model of a hydrated dimyristoylphosphatidylcholine bilayer.

(A) Icon of the lipid bilayer based on the model shown in (B). (B) Space-filling model of all the lipid atoms in the simulation. Stick figures of the water molecules are red. The polar regions of phosphatidylcholine (PC) from the carbonyl oxygen to the choline nitrogen are blue. Hydrocarbon tails are yellow. (C) Stick figures of the water molecules only. (D) Stick figures of the polar regions of PC from the carbonyl oxygen to the choline nitrogen only. (E) Stick figures of the hydrocarbon tails only. This model was calculated from first principles starting with 100 PC molecules (based on an X-ray diffraction structure of PC crystals) in a regular bilayer with 1050 molecules of bulk phase water on each side. Taking into account surface tension and distribution of charge on lipid and water, the computer used simple Newtonian mechanics to simulate the molecular motion of all atoms on a picosecond time scale. After less than 100 ps of simulated time, the liquid phase of the lipids appeared. The model shown here is after 300 ps of simulated time. Such models account for most molecular parameters [electron density, surface roughness, distance between phosphates of the two halves, area per lipid ($0.6\,nm^2$), and depth of water penetration] of similar bilayers obtained by averaging techniques, including nuclear magnetic resonance, X-ray diffraction, and neutron diffraction.

Pollard TD, Earnshaw WC, Lippincott-Schwartz J, Johnson GT. Cell Biology. 3rd ed. Philadelphia, PA: Elsevier; 2017. Fig. 13.5, p. 231. With permission.

catalytic site for peptide bond formation is composed of RNA, which is surrounded by numerous proteins. The donors of the amino acids that are added to the peptide chain from N terminal to C terminal are also composed of a small transfer or tRNA attached to an amino acid.

At a somewhat later date, proteins became a key component of early cells. There are many advantages to proteins compared with nucleic acids in carrying out the processes necessary for life, including energy production, metabolism, and the synthesis of proteins, nucleic acids, phospholipids, and long chain sugar compounds. Proteins are constructed from amino acids. The structures, names, abbreviations, and polarities (i.e., the solubility in water) of the 20 amino acids are shown in Fig. 2.2. Because there are 20 amino acids compared with only four nucleotides, proteins can form many more individual molecules. As described in Chapter 1, Classes of receptors, their signaling pathways, and their synthesis and transport, proteins are modular and can be assembled into an almost infinite variety of molecules. It seems as though once a useful module evolved, it was used for the construction of many different proteins during the course of evolution.

First position (5′ end)	Second position: U	Second position: C	Second position: A	Second position: G	Third position (3′ end)
U	UUU, UUC – Phe UUA, UUG – Leu	UCU, UCC, UCA, UCG – Ser	UAU, UAC – Tyr ● UAA, ● UAG – Stop	UGU, UGC – Cys ● UGA Stop UGG Trp	U C A G
C	CUU, CUC, CUA, CUG – Leu	CCU, CCC, CCA, CCG – Pro	CAU, CAC – His CAA, CAG – Gln	CGU, CGC, CGA, CGG – Arg	U C A G
A	AUU, AUC, AUA – Ile ○ AUG Met	ACU, ACC, ACA, ACG – Thr	AAU, AAC – Asn AAA, AAG – Lys	AGU, AGC – Ser AGA, AGG – Arg	U C A G
G	GUU, GUC, GUA, GUG – Val	GCU, GCC, GCA, GCG – Ala	GAU, GAC – Asp GAA, GAG – Glu	GGU, GGC, GGA, GGG – Gly	U C A G

● = Chain-terminating codon
○ = Initiation codon

FIGURE 2.2 The genetic code.

The locations of the nucleotide in first, second, and third positions define the amino acid specified by the code.

Pollard TD, Earnshaw WC, Lippincott-Schwartz J, Johnson GT. Cell Biology. 3rd ed. Philadelphia, PA: Elsevier; 2017. Fig. 12.2, p. 210. With permission.

These modules can be thought of as individual LEGOS that can be assembled into a myriad of different shapes with different functions.

As discussed in Chapter 1, Classes of receptors, their signaling pathways, and their synthesis and transport, proteins can exist both in the watery cytoplasm of cells, or can be partially or almost fully embedded in the cells' hydrophobic ("water hating") membranes. Also as discussed in Chapter 1, Classes of receptors, their signaling pathways, and their synthesis and transport, proteins have the ability to change their shape when they encounter another protein or small molecule. This makes them very useful as receptors for an almost infinite number of small and large molecules in their environment. Besides their unique ability to function as receptors, another critical function of many proteins is to serve as enzymes that catalyze a huge variety of chemical reactions.

Scientists have recently sequenced the genome of the last universal ancestor of life (LUCA). This one-celled organism, a member of the most ancient phylum, Archaea, has only 355 protein and RNA coding genes. LUCA derived its energy from nutrients in the hot vents on the sea floor.

Also very recently researchers have synthesized a viable single-celled organism by minimization of the genome of a Mycoplasma. It contains 473 genes, a smaller number than that in *Mycoplasma genitalium*. Interestingly, this synthetic genome contains 149 genes of completely unknown function, suggesting that there are as yet undiscovered functions essential for life.

2.2 ROLES OF RECEPTORS AND THEIR COGNATE LIGANDS IN EARLY ARCHAEA AND BACTERIAL EVOLUTION

Three and a half billion years ago, single-celled organisms quite similar to those found today were present. Stromatolites, consisting of many sheets of fossilized Cyanobacteria, often called blue green algae, have been found in Australia as well as in other parts of the world. Blue green algae are members of the most ancient phylum, Archaea, the ancestor of all other living organisms.

Blue green algae played a key role in the evolution of higher plants and animals because they were able to convert CO_2 in their chloroplasts into complex carbon-containing molecules. This process liberated O_2 into the atmosphere. The O_2-containing atmosphere in turn was essential for the evolution of all eukaryotes.

Archaea and Bacteria, non-nucleus-containing organisms, need to carry out two major functions: to replicate and to gain information about their environment. The latter allowed them to avoid prey and to find food. The keys to information gathering, that is, receptors, are also critical in the long course of evolution leading to the most complex organ, the mammalian brain.

As was discussed in Chapter 1, Classes of receptors, their signaling pathways, and their synthesis and transport, receptors are proteins that can very sensitively recognize specific molecules, and in most cases, transmit a signal upon binding.

Archaea contain the Che receptor similar to that in *Escherichia coli* and other bacterial species. The family of Che receptors in *E. coli* consists of four members, each of which can specifically and sensitively bind to a small molecule near their cell surface that can serve as food. The Tar receptor recognizes one sugar, maltose, and the amino acid, Asn. The Tsr receptor recognizes the amino acids Ser and Leu. The Trg receptor recognizes two sugars, ribose and galactose, and the Tap receptor binds to dipeptides derived from proteins, and to pyrimidines derived from the nucleic acids, cytosine, and thymidine.

All four receptors have similar structures, an N-terminal region that protrudes from the extracellular side of the plasma membrane, a middle region that traverses the plasma membrane, and a C-terminal region that serves as a signaling domain and contains a methyl group–accepting domain.

When a particular receptor binds to its cognate ligand, the receptor changes its shape. In the case of a Che receptor, the signaling domain is now recognized by several proteins, including the CheA sensor kinase and the CheY adapter protein as well as by a methylating enzyme, CheW. The CheA adds phosphate residues to CheY at a decreased rate, which causes the bacterial flagella to turn the bacterium toward the food source. The signal can also be modulated by CheY, the methylating enzyme, as well as by enzymes that remove the phosphate and methyl groups respectively. Other receptors recognize ligands that indicate a toxic environment. These receptors signal to the flagella to reverse course to escape the noxious stimulus using a similar mechanism.

It is obvious from these examples that receptors, with their ability to very sensitively recognize specific molecules in the environment and signal their presence to the cell, constitute an incredibly powerful system that has served all organisms throughout evolutionary history.

Another class of receptors evolved at the dawn of evolution. These are membrane proteins that traverse the plasma membrane seven times and form a small hydrophilic ("water loving") selective channel in the middle. As was discussed in Chapter 1, Classes of receptors, their signaling pathways, and their synthesis and transport, a key member of this very important receptor class (and the first membrane protein whose structure was established by X-ray crystallography) is Bacteriorhodopsin from the Archaea *Halobacterium halobium*. This molecule has an attached retinal (vitamin A aldehyde) that changes from one conformation to another (*cis* to *trans*) when a photon of light strikes it (see Fig. 1.3). This conformational change opens a selective channel for H+ ions, which are then pumped out of the cell, producing a concentration gradient, which generates energy that is used by the cell to make ATP molecules.

Halobacterium contains another evolutionarily similar molecule, the receptor Halorhodopsin. This protein has evolved to fulfill a totally different function: phototaxis or light seeking. Halorhodopsin couples light absorption by the bound retinal to a conformational change in the molecule, enabling it to bind a signaling protein, which produces phototaxis ("swimming to the light"), a necessity for this photosynthetic organism.

Both of these rhodopsins are structurally and functionally quite similar to the vertebrate rhodopsins, the visual receptors of the photosensory cells of the retina, even though they don't appear to be evolutionarily related. The vertebrate rhodopsins in photoreceptor cells interact with G proteins. This in turn transmits the signal to other cells in the retina and ultimately into the brain, allowing vision to occur. As was discussed in Chapter 1, Classes of receptors, their signaling pathways, and their synthesis and transport, members of the seven membrane-spanning receptor superfamily, the G-protein coupled receptors (GPCRs), have evolved to serve as receptors for hormones, neurotransmitters, odors, pheromones, etc. GPCRs form the most abundant class of molecules found in the vertebrate genome.

In bacteria, receptors also play key roles in the process called quorum sensing. Research has demonstrated that conversion to a so-called competent state in *Streptococcus* is governed by an extracellular factor manufactured by *Streptococcus* itself. The competence factor, which was later shown to be a modified peptide, was a "hormone-like activator" that synchronizes the behavior of the bacterial population by binding to a specific receptor. In 1970, two species of bioluminescent marine bacteria, *Vibrio fischeri* and *Vibrio harveyi*, were shown to produce light when cultured at high cell density but not in dilute suspensions. Light production could be stimulated by the exogenous addition of cell-free culture fluids, and the responsible molecule, called the autoinducer, was later identified as an acyl-homoserine lactone (AHL). Again it functioned by binding to its specific receptor. These combined results suggest that certain bacteria use the production, release, exchange, and detection of signaling molecules to measure their population density. This receptor-mediated process allows the bacteria to control their behavior in response to variations in cell numbers. It is now clear that quorum sensing is not the exception. Rather, it is very common behavior in the bacterial world and this process is fundamental to all of microbiology.

How does quorum sensing work? As a population of quorum-sensing bacteria grows, a proportional increase in the extracellular concentration of the signaling molecule occurs. When a threshold concentration is reached, the group detects the signaling molecule via its binding to its cognate receptors, and responds to it with a population-wide alteration in gene expression. Processes controlled by quorum sensing tend to be those that are unproductive when performed by an individual bacterium, but that become effective when carried out by the group. For example, in addition to competence and bioluminescence, quorum sensing controls virulence factor secretion, biofilm formation, and sporulation. Thus quorum sensing is a mechanism that allows bacteria to function as multicellular organisms and to reap benefits that they could never obtain if they only existed as single cells.

Gram-positive quorum-sensing bacteria such as *Streptococcus* and *Bacillus* communicate with short peptides that often contain chemical modifications.

The signaling molecule for genetic competence in *B. subtilis*, ComX, is a 6—amino acid peptide whose tryptophan residue has been modified by the attachment of a geranyl group. Signaling peptides such as ComX are recognized by membrane-bound two-component receptor-sensing histidine kinases. Signal transduction occurs by phosphorylation cascades that ultimately converge on DNA binding transcription factors responsible for regulation of target genes. In general, bacteria keep their AHLs and peptide quorum-sensing conversations private. This is accomplished by each species of bacteria producing and detecting a unique AHL. AHLs differ in their acyl side-chain moieties, peptides, or combinations thereof.

Many bacteria are also capable of coordinating their behavior to form complex multicellular communities consisting of large numbers of densely packed cells. These architecturally complex communities, called biofilms, form on surfaces or at air—liquid interfaces. Cells in biofilms are held together by an extracellular matrix composed of polysaccharides, protein, and often DNA. All of the members of a bacterial community cooperate in the construction of the biofilm by contributing matrix components. These extracellular matrix components, which have structural and functional similarities to those found in eukaryotic multicellular organisms, form a protective coat around the colony and communicate with the bacterial cells via receptors on their cell surfaces. The matrix is thought to help render the colony resistant to antibiotics; hence the great importance of biofilms in human diseases such as tuberculosis. Also each individual bacterium within the biofilm expresses distinct metabolic pathways, stress responses, and other biological activities. Thus the biofilm bestows many of the benefits of multicellularity possessed by true multicellular organisms.

Scientists recently analyzed the DNA of an Archaea living in an undersea vent. It turned out that the sediment contained DNA from a lineage of Archaea unlike any previously discovered. The scientists dubbed it Lokiarchaeum, named for a hydrothermal vent called Loki's Castle near the location where the Archaea were found.

This DNA is far more closely related to eukaryotes than any other known species of Archaea. It also contains genes for many traits that previously were only known to be present in eukaryotes. Among these genes were many that are components of a specialized organelle inside eukaryotic cells, the lysosome. Inside this compartment eukaryotic cells can destroy defective proteins and proteins engulfed from outside the cell.

All eukaryotes also share a cellular skeleton, which they constantly construct and destroy to modify their shape. Many genes in Lokiarchaeum encode the proteins required to carry out these tasks. It's possible that Lokiarchaeum uses its skeleton to crawl over surfaces as protozoans do. Lokiarchaeum's genes also suggest that it can swallow up molecules or microbes as eukaryotes do.

All in all, Lokiarchaeum is much more complex than other Archaea and bacteria, although not as complex as true eukaryotes. It lacks a nucleus and mitochondria. Therefore it can be thought to be an evolutionary precursor to all eukaryotes.

2.3 CHOANOFLAGELLATES AND THE EVOLUTION OF G-PROTEIN COUPLED RECEPTORS

There is now substantial evidence that the closest living organisms sharing a common ancestor with all multicellular organisms are Choanoflagellates. These protists each possess a single long flagellum that beats water. This in turn brings assorted bacteria to their cell surface. Then the bacteria are engulfed by a process called phagocytosis into the interior of the cell where they are digested. Many multicellular organisms, including sponges and Hydra, have similar cells to those of the Choanoflagellates called collar cells. Collar cells have an identical function, that is, to use their long flagella to beat, thereby aiding in feeding. Therefore they may have evolved from the Choanoflagellates.

One mystery regarding the Choanoflagellate digestive system has very recently been solved. It has been discovered that the organism has a digestive system as well as an anal sphincter.

The genomes of two Choanoflagellate species have recently been sequenced and they contain a number of sequences that are homologous to those found in multicellular organisms. Two proteins that are homologous to receptors involved in brain development are Notch and cadherin.

Another group of important receptors are present in Choanoflagellates: the GPCRs. As mentioned in Chapter 1, Classes of receptors, their signaling pathways, and their synthesis and transport, GPCRs are the largest and most important group of receptors in the mammalian genome. There are over 800 of these receptors in the human genome, and more than half of the drugs on the market are directed at GPCRs. They are also predominantly expressed in the brain and many have been shown to be involved in developmental processes.

In contrast to the 10 GPCR genes in Choanoflagellates, there was a huge expansion of the GPCR genes in multicellular organisms, including sponges and comb jellies, which don't contain nervous systems, and Hydra and other nonbilateral multicellular organisms, which do. For example, sponges contain more than 330 GPCRs and Hydra over 1200. These findings indicate that there was a large expansion with the advent of multicellularity and another significant expansion with the evolution of nerve cells.

As we have discussed in Chapter 1, Classes of receptors, their signaling pathways, and their synthesis and transport, and this chapter, the GPCRs are an ancient family, structurally and functionally related to proteins found in the earliest Archaea. The functions of these proteins, Bacteriorhodopsin and Halorhodopsin, either produce ATP by pumping H^+ out of the cell, or sense the presence and direction of light. While there is no conclusive evidence for their evolutionary similarity with the GPCRs found in eukaryotes, recent work has shown that there is a strong similarity between the hydrophilic amino acids buried in the membranes of the Archaea proteins and those in the GPCRs. This discovery makes it likely that they evolved from a common ancestor.

It is thought that the GPCRs evolved from a single ancestral protein by two distinct mechanisms. One involves the duplication of a single region of a chromosome containing a GPCR gene. The other involves the duplication of an entire chromosome.

There are five families of GPCRs that evolved quite early in evolution: the glutamate family, the rhodopsin family, the adhesion family, the frizzled family, and the secretin family. These families together are known by the acronym the GRAFS families. The Rhodopsin family is by far the most abundant family, consisting of over 800 odorant receptors found in nasal epithelial cells, as well as receptors for a number of neuropeptides and neurotransmitters. The Glutamate family is composed of receptors for the excitatory transmitter Glutamate and those for the inhibitory transmitter GABA. This family also includes several taste receptors. The adhesion family contains receptors with long N-terminal domains, which can make contacts with adhesion molecules on other cells. Several members of this family have been shown to be involved in neural development. The Frizzled group contains several receptors for Wnt, a peptide that plays a key role in neuronal development. Also this family contains several additional taste receptors. The Secretin family contains a number of important neuropeptide and hormone receptors.

As discussed in Chapter 1, Classes of receptors, their signaling pathways, and their synthesis and transport, the GPCRs signal through three protein containing complexes, called G proteins. These three components of all G proteins are $G\alpha$, $G\beta$, and $G\gamma$. The G proteins are also an ancient lineage but have evolved separately from the GPCRs. There are two orders of magnitude fewer genes coding for G proteins than there are GPCR genes in multicellular organisms.

RECOMMENDED BOOK CHAPTERS

Lodish H, Berk A, Kaiser CA, et al. *Molecular Cell Biology*. 8th ed. New York: W.H. Freeman; 2016 [Chapters 1, 17].

Pollard TD, Earnshaw WC, Lippincott-Schwartz J, Johnson GT. *Cell Biology*. 3rd ed. Philadelphia, PA: Elsevier; 2017 [Chapters 2, 22, 27].

Review Articles

Adamska M. Sponges as models to study emergence of complex animals. *Curr Opin Genet Dev*. 2016;39:21−28.

Altman S. Nobel lecture. Enzymatic cleavage of RNA by RNA. *Biosci Rep*. 1990;10 (4):317−337.

Cavalier-Smith T. Origin of animal multicellularity: precursors, causes, consequences-the choanoflagellate/sponge transition, neurogenesis and the Cambrian explosion. *Philos Trans R Soc Lond B Biol Sci*. 2017;372:1713−1717.

Flemming H, Wingender J, Szewzyk U, Steinberg P, Rice S, Kjelleberg S. Biofilms: an emergent form of bacterial life. *Nat Rev Microbiol*. 2016;14:563−575.

Koshland DE. Chemotaxis as a model second-messenger system. *Biochemistry*. 1988;27:5829−5834.

Krishnan A, Schiöth HB. The role of G protein-coupled receptors in the early evolution of neurotransmission and the nervous system. *J Exp Biol*. 2015;218:562–571.

Orgel LE. Prebiotic chemistry and the origin of the RNA world. *Crit Rev Biochem Mol Biol*. 2004;39:99–123.

Schrum JP, Zhu TF, Szostak JW. The origins of cellular life. *Cold Spring Harb Perspect Biol*. 2010;. a002212.

Dictyostelium discoideum, sponges, comb jellies, and hydra: the earliest animals

3

CHAPTER OUTLINE

3.1 DICTYOSTELIUM

While it is true that some classes of bacteria and other protists can form multicellular and differentiated complexes with more than one cell type, the most evolutionarily relevant organism for the eukaryotic transition from a single cell to a multicellular organism is the cellular slime mold *Dictyostelium discoideum*. This organism is a member of the same class as yeast and other fungi.

During most of its life cycle, *Dictyostelium* exists as a single amoeboid cell, moving through the soil and feeding on all types of bacteria. When food becomes scarce, however, individual cells aggregate by a chemotropic process, following a gradient of extracellular cAMP secreted by certain cells. The key protein involved in this excellent example of tropism, that is, the migration of cells along a positive gradient toward the source of a substance, is the cAMP receptor.

How do the migrating amoebas respond to the cAMP gradient? Scientists have examined the various components involved and demonstrated that both cAMP receptors and their trimeric G proteins are evenly distributed throughout the cell membrane. However, the enzyme phosphoinositol-3 kinase (PI-3 kinase), which produces the signaling molecule, PI 3,4,5 triphosphate, is distributed in a gradient with more being present at the front of the cell than at the rear. This skewed distribution occurs after the cells are exposed to a gradient of cAMP. Proteins involved in cell motility, including those essential for the polymerization of actin filaments, part of the motile apparatus, also become localized at the forward end of the migrating cell. Signaling via the cAMP receptor triggers these changes. This system serves as a model for chemotactic cells in higher organisms such as microglia in the brain. Other G-protein coupled receptors (GPCRs) are important in the differentiation of the slug, including Class C GPCRs that bind to GABA and Glu and signal through different trimeric G proteins.

Receptors in the Evolution and Development of the Brain. DOI: https://doi.org/10.1016/B978-0-12-811012-6.00003-0

After migration is complete the cells form a slug, mainly composed of two types of cells: stalk forming cells (80%) and spore forming cells (20%). The latter function as the reproductive cells. After completion of the slug, which is now referred to as the fruiting body, the spores are dispersed by a variety of methods including high winds. Once the spores have landed in a new location, they remain dormant until they sense new sources of food and become amoeba once more.

Including the genes coding for the proteins discussed above, *D. discoideum* contains 13,498 genes—about two-thirds the number contained in the human genome. The *Dictyostelium* genes are also more similar to those of humans than to those of yeast, another member of the fungus family. Its simple differentiation system, combined with the similarity of its genome to that of humans, has made it an excellent model system to study cellular differentiation.

3.2 SPONGES

Sponges are one of the most ancient multicellular animals. Sponges are animals in the phylum Porifera, whose name comes from the Greek word meaning "pore bearer." They evolved more than 840 million years ago, and therefore can be thought of as living fossils. There are several types of sponges, and many scientists used to believe that one type of sponge was a direct ancestor of all higher animals with nervous systems, but now this is not thought to be the case. Rather it is believed that sponges are descended from animals possessing a nervous system, but that was was dispensed with during further evolution.

Sponges are multicellular animals composed of two distinct layers of cells, and between them is a jelly-like mesohyl. Sponges do not have digestive, circulatory, or nervous systems. Rather, most sponge species rely on maintaining a constant flow of water through the channels and pores comprising their bodies. This movement allows them to maintain a constant flow of oxygen and food and to remove waste. They employ collar cells, which beat using their flagellar apparatus to force water through their pores. These collar cells are strikingly similar in both structure and function to single-celled Choanoflagellates.

Even though sponges do not have nervous systems, the recent determination of the complete genetic code of the sponge *Amphimedon queenslandia* has demonstrated that they have DNA sequences that code for essentially all the proteins that are found in the postsynaptic density, an evolutionarily conserved and critical element of the postsynaptic ending of neurons. These sequences have been shown to be transcribed into mRNAs. The expression of the postsynaptic density genes has been shown to occur in larval flask cells, strongly suggesting that these gene products have a critical but as yet undetermined function in sponge development (see Chapter 9: The importance of rapid eye movement sleep and other forms of sleep in selecting the appropriate neuronal circuitry; programmed cell death/apoptosis). Genes coding for proteins that are present in presynaptic endings are also found in sponges. At present it is not known if these presynaptic gene sequences are translated into proteins, and if so, where these proteins are located. Obviously it will be of great significance to determine the function(s) of these proteins.

The recent genomic sequencing of *A. queenslandia* has also revealed that many genes involved in nerve cell development, including members of the Wnt, TGFβ, Notch, and Hedgehog signaling pathways, and the receptors associated with them, are present. Specific transcription factors associated with nervous system development, including Sox B, Mef2, Irx, and BHLH, are also found in sponges. Proteins associated with programmed cell death or apoptosis, which is crucial to some vertebrates' nervous system development, are found in the sponge genome, including those of initiator caspases. Finally, members of two major cell—cell adhesion superfamilies, those containing the cadherin domain and those containing the immunoglobulin domain—both of which play key roles in nervous system development—are also found in the sponge. These topics will be discussed in greater detail in later chapters.

While there are no nerve cells in sponges, the flask cells found in the sponge larvae contain rhodopsins, members of the GPCR superfamily, which allow them to sense light. These cells are embedded within an epithelial layer of polarized ciliated cells. Therefore the flask cells may transmit the sensory information they obtain to the ciliated cells through the secretion of undetermined paracrine factors, causing them to adjust the direction in which they swim. This combination of flask and ciliated cell types may be thought of as homologous to a sensory cell—muscle cell combination with no intermediary central nervous system.

It is now believed that the direct ancestor of animals with nervous systems bore a close similarity to Ctenophores, the comb jellies. These predators move by means of ciliated cells on their outer epithelium. They have an internal digestive system with a mouth, gut, and anus. Evidence to date indicates that the comb jellies have no true nervous system, instead possessing a network of interconnecting neurons called a nerve net.

3.3 HYDRAS

Hydras and other Cnidaria, including jellyfish, sea anemones, and corals, are among the first animals with true but simple nervous systems. Hydras were first observed with a microscope in 1704 and represent one of the most primitive members of Cnidaria. Hydras have only two layers of cells, ectoderm and endoderm, separated by an extracellular matrix, the mesoglea. They consist of a head with tentacles to seize their prey, a gastric column, and a foot. It was thought for many years that Hydras had a very simple nervous system consisting of a nerve net. In recent years, however, Hydras have been shown to have a much more complex nervous system consisting of both ganglion cells, which function as interneurons, and sensory neurons. They also have neurons that form chemical synapses with myoepithelial cells. The presynaptic endings of these neurons contain dense core synaptic vesicles containing one or more neuropeptides, as well as the machinery for the production of acetylcholine, a neurotransmitter employed at the neuromuscular synapse of higher animals including humans. Therefore these neurons can be thought of as an evolutionary precursor to motor neurons. The structure of the vertebrate synapse will be discussed in detail in Chapter 4, The growth cone: the key organelle in the steering of the neuronal processes; the synapse: role in neurotransmitter release.

The Hydra's neurons differentiate from a class of cells called the interstitial cells that are found in the mesoglial layer and are stem cells, meaning they can differentiate into any cell type in the Hydra's body. These include cnidocytes, the stinging cells of the Hydra, as well as gland cells and gametes. These stem cells are found throughout the gastric column but not in either the head or the foot. Upon differentiating into neurons, the cells migrate to the appropriate place in the Hydra. These stem cells may be evolutionary precursors of the stem cells found in the developing and mature vertebrate brain and other organs. Stem cells in the mammalian brain will be discussed in Chapter 19, Potential use of neuronal stem cells to replace dying neurons depends on trophic factors and receptors.

Hydras contain many small peptides, perhaps more than 1000. This is almost an order of magnitude greater than those found in humans. Some of these peptides localize to neurons, others to epithelial cells. At least 850 different peptides have been isolated from Hydra and many of these have been sequenced. As many as 47 copies of a single neuropeptide have been found in a single propeptide precursor.

These neuropeptides are very likely keys to the development of the Hydra and also serve as the major neurotransmitters. Since neuropeptides mainly signal through GPCRs, they require seconds or more to perform their signaling functions. Therefore it is necessary for some neuropeptides to activate ion channels, leading to very rapid signaling. At least two of these channel-activating neuropeptides have been characterized from Hydras and it is likely that many more exist.

As mentioned, mammals and other vertebrates contain many fewer neuropeptides. This is likely because they use small molecules such as Glu and GABA as the major neurotransmitters and therefore require fewer neuropeptides. As will be discussed in Chapter 5, Neuronal cell biology, small neurotransmitters can be used many times without resynthesis while neuropeptides can only be used once. Therefore as neurons became longer they required molecules that can be replenished locally to be used as neurotransmitters, while neuropeptides are used as neuromodulators. Neuropeptides are also used as neurohormones that are released into the blood. This subject will be described at length in Chapter 13, Steroid hormones and their receptors are key to sexual differentiation of the brain while glucocorticoids and their receptors are key components of brain development and stress tolerance; the hypothalamic hormone producing cells: importance in various functions during and after brain development.

Few receptors for these Hydra peptides have been identified, except for one interesting peptide, the "head activator" discussed in the following paragraph. The genome of a Hydra has also quite recently been sequenced, allowing for the eventual determination of receptor number in this "simple" animal. Despite the apparent simplicity of Hydra, its genome contains over 20,000 genes, approximately the same number of genes found in humans.

The head activator peptide is an 11−amino acid−containing neuropeptide purified from Hydras undergoing head regeneration. The head activator stimulates neuronal stem cell division, neuronal differentiation, and migration. Homologues are found in all animals, including humans and insects, and the head activator has a positive effect on cell division and neuronal differentiation in all species studied.

Two candidate receptors for the head activator have been isolated. One receptor is a single membrane-spanning protein with most of its amino acids on the extracellular side. It is called Sortilin and is found in all animal species. Sortillin's extracellular domain contains a region called the vps10 domain. This domain recognizes many cargo proteins in a late endosome and shuttles them to the Golgi apparatus. The head activator binds to the vps10 domain with high affinity. Another receptor called GPCR37 has also been found to bind to the head activator. Binding of head receptor to GPCR37 activates cell division through an inhibitory G-protein linked pathway, by activation of Ca^{++} influx. It is not yet known if this receptor is found in Hydras.

Beside the peptides and receptors described above, Hydras also contains master gene products that control the differentiation of the nervous system as well as specification of the body plan. These include the signaling protein, Wnt, and its plasma membrane receptor, Frizzled. Finally, Hydras also contain genes specifying transcription factors that control the expression of key proteins involved in the differentiation of the nervous system.

RECOMMENDED BOOK CHAPTERS

Lodish H, Berk A, Kaiser CA, et al. *Molecular Cell Biology*. 8th ed. New York: W.H. Freeman; 2016 [Chapter 17].

Pollard TD, Earnshaw WC, Lippincott-Schwartz J, Johnson GT. *Cell Biology*. 3rd ed. Philadelphia, PA: Elsevier; 2017 [Chapters 26, 38].

Review Articles

Cavalier-Smith T. Origin of animal multicellularity: precursors, causes, consequences-the choanoflagellate/sponge transition, neurogenesis and the Cambrian explosion. *Philos Trans R Soc Lond B Biol Sci*. 2017;372:1713−1717.

Devreotes P. *Dictyostelium discoideum*: a model system for cell-cell interactions in development. *Science*. 1989;245:1054−1058.

Krishnan A, Schiöth HB. The role of G protein-coupled receptors in the early evolution of neurotransmission and the nervous system. *J Exp Biol*. 2015;218:562−571.

Rappel WJ, Loomis WF. Eukaryotic chemotaxis. *Wiley Interdiscip Rev Syst Biol Med.* 2009;1:141−149.

Takahashi T, Takeda N. Insight into the molecular and functional diversity of cnidarian neuropeptides. *Int J Mol Sci.* 2015;16:2610−2625.

The growth cone and the synapse

4

CHAPTER OUTLINE

4.1 GROWTH CONE

A key step in the evolution of a complex nervous system was the development of the neuronal growth cone. Initially each developing neuron must wend its way through a dense jungle of other neurons as it puts out processes that are destined to become one axon and one or more dendrites. For this task, the neuron uses receptors on its cell surface, many expressed at the process endings called growth cones (Fig. 4.1). These receptors recognize complementary ligands, either attractive or repulsive, and convey this information to the growth cones, which are constantly integrating a whole host of signals and deciding in which direction to move (Fig. 4.2). Therefore, growth cones can be considered the brain of the growing neuron.

The attractive molecules are called tropic molecules. They may or may not also be trophic molecules. There are also repulsive signals as discussed below. In concert, these signals allow the growth cone to grow in a precise direction toward its final destination.

The growth cone membrane consists of a number of long, narrow motile elements called filopodia and between them, ruffled patches called lamellipodia (Fig. 4.3). These two elements move the neurite via the parallel filaments comprised of the protein actin, which is almost identical in both structure and function to the thin filament contractile protein of skeletal muscles. All actin filaments are attached at one end to the plasma membrane by interactions with a complex of peripheral and integral membrane proteins (Fig. 4.3). In the filopodia, parallel actin filaments are attached to the far end of the filopodia and can polymerize at the membrane tip and depolymerize at the base (Fig. 4.4). Actin is bound to the protein tropomyosin, an elongated α-helical coiled coil that runs parallel to the double helical actin filament and lies in its groove. Tropomyosin and several

Receptors in the Evolution and Development of the Brain. DOI: https://doi.org/10.1016/B978-0-12-811012-6.00004-2

51

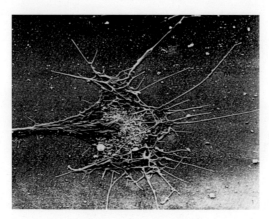

FIGURE 4.1 Electron micrograph of a growth cone in culture.

Growth cones extending on a flat surface are typically very thin, with broad lamellae and numerous filopodia.

Squire LR, Berg D, Bloom FB, Du Lac S, Ghosh A, Spitzer NC. Fundamental Neuroscience. 4th ed. Oxford: Elsevier; 2013 [Chapters 6–9, 16]. Fig. 16.1, p. 364. With permission.

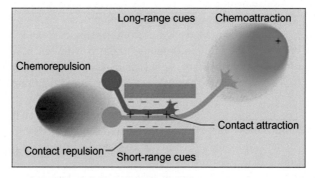

FIGURE 4.2 Axons are guided by the simultaneous and coordinate actions of four types of guidance mechanisms: contact attraction, chemoattraction, contact repulsion, and chemorepulsion.

Individual growth cones might be "pushed" from behind by a chemorepellent, "pulled" from in front by a chemoattractant, and "hemmed in" by attractive and repulsive local cues (cell surface or extracellular matrix molecules). Push, pull, and hem: these forces act together to ensure accurate guidance.

Squire LR, Berg D, Bloom FB, Du Lac S, Ghosh A, Spitzer NC. Fundamental Neuroscience. 4th ed. Oxford: Elsevier; 2013 [Chapters 6–9, 16]. Fig. 16.4, p. 366. With permission.

other proteins control the polymerization of actin. The motor protein, myosin, an ATPase, transports membrane vesicles along the parallel actin filaments to the tips of the filopodia (Fig. 4.5). The neurite grows by fusion of the membrane vesicles at these tips (Fig. 4.3).

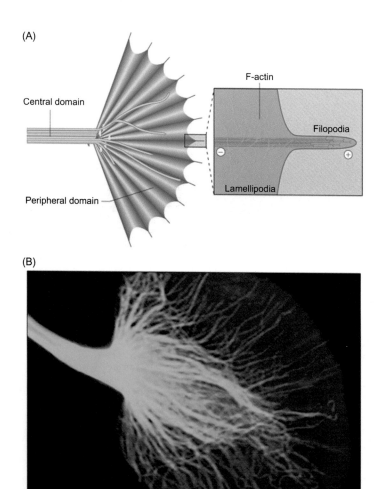

FIGURE 4.3 Distributions of microtubules and fibrillary actin in a neuronal growth cone.

(A) Microtubules and fibrillary actin are highly compartmentalized within growth cones. Microtubules (*green*) are a key structural component of the axon and though tightly bundled with the axon shaft, a dynamic subpopulation of microtubules actively explores the peripheral domain of the growth cone by extending along filopodia. All of the growing ends of microtubules are pointed toward the leading edge. In contrast, actin (*red*) is highly concentrated in the filopodia and in the leading edges of lamellae. Within filopodia, actin fibrils are oriented with their growing tips pointed distally. The same is true of many fibrils within lamellae, although many additional fibrils are oriented randomly and form a dense meshwork. (B) The organization of cytoplasmic domains and cytoskeletal components in an Aplysia bag cell growth cone. Fluorescent labeling of microtubules (*green*) and fibrillar actin (*red*) shows their segregation within the growth cone central and peripheral domains, respectively. Note the extension of a small number of microtubules into the actin-rich peripheral domain.

Squire LR, Berg D, Bloom FB, Du Lac S, Ghosh A, Spitzer NC. Fundamental Neuroscience. *4th ed. Oxford: Elsevier; 2013 [Chapters 6–9, 16]. Fig. 16.9, p. 375. With permission.*

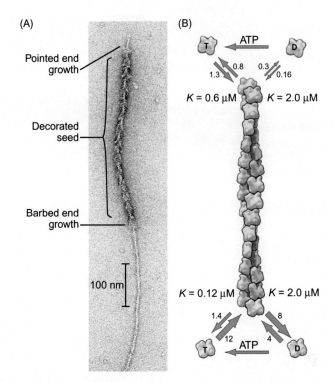

FIGURE 4.4 Actin filament elongation.

(A) Electron micrograph of growth from an actin filament "seed" decorated with myosin heads to reveal the polarity. Growth is faster at the barbed end than at the pointed end. (B) Rate constants for association (units: $\mu M^{-1} s^{-1}$) and dissociation (units: s^{-1}) for Mg-ATP-actin (T) and Mg-ADP-actin (D) were determined by measuring the rate of elongation at the two ends as a function of monomer concentration. Ratios of the rate constants yield critical concentrations (*K*, units: μM) for each reaction. Critical concentrations at the two ends are the same for ADP-actin but differ for ATP-actin. *ADP*, adenosine diphosphate.

Pollard TD, Earnshaw WC, Lippincott-Schwartz J, Johnson GT. Cell Biology. 3rd ed. Philadelphia, PA: Elsevier; 2017 [Chapters 1, 3, 13–16, 21–24, 27, 30, 33, and 34]. Fig. 33.8, p. 579. With permission.

Key proteins that control the state of the actin filaments are small GTPases that are members of the Rho-family. These proteins, called Cdc42, Rac, and Rho, are coordinately activated by the binding of an attractive ligand to its receptor and produce a change in the direction of growth (Fig. 4.6). The higher concentration of tropic molecules on one side of the filopodia causes the preferential formation of actin filaments on this side of the growth cone, producing a shift in the direction of growth. An illustration of the effect of either attractive or repulsive molecules is shown in Fig. 4.7. Guidance molecules shown will be discussed in detail in Chapter 11, The key roles of axon and dendrite guidance molecules and the extracellular matrix in all stages of brain development.

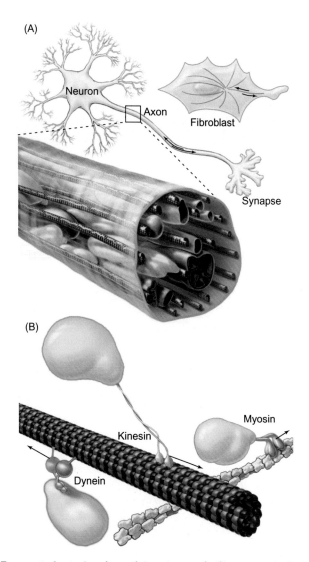

FIGURE 4.5 Transport of cytoplasmic particles along actin filaments and microtubules by motor proteins.

(A) Overview of organelle movements in a neuron and fibroblast. (B) Details of the molecular motors. The microtubule-based motors, dynein and kinesin, move in opposite directions. The actin- based motor, myosin, moves in one direction along actin laments.

Pollard TD, Earnshaw WC, Lippincott-Schwartz J, Johnson GT. Cell Biology. 3rd ed. Philadelphia, PA: Elsevier; 2017 [Chapters 1, 3, 13–16, 21–24, 27, 30, 33, and 34]. Fig. 1.14, p. 13. With permission.

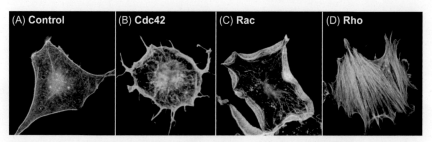

FIGURE 4.6 Rho-family GTPases promote the assembly of actin-based structures.

Fluorescence micrographs of Swiss 3T3 fibroblasts stained with rhodamine-phalloidin to reveal actin filaments. (A) Resting cells. (B) Cells microinjected with activated Cdc42 form many filopodia. (C) Cells microinjected with activated Rac have a thick cortical network of actin laments around the periphery. (D) Stress fibers anchored at their ends by focal contacts are abundant in cells microinjected with an activated form of Rho.

Pollard TD, Earnshaw WC, Lippincott-Schwartz J, Johnson GT. Cell Biology. 3rd ed. Philadelphia, PA: Elsevier; 2017 [Chapters 1, 3, 13–16, 21–24, 27, 30, 33, and 34]. Fig. 33.19, p. 588. With permission.

Underlying the tip of the membrane is a core of microtubules and mitochondria. The molecular structure of a microtubule and its subunit, tubulin, is shown in Fig. 4.8. The microtubules provide stability to the long, narrow neurite. They also stabilize the filopodia during process growth by providing one or more microtubules that grow into each filopodium. The mitochondria provide a ready source of energy to the metabolically extremely active growth cone.

As mentioned previously, the growing nerve process elongates by membrane accretion at the tip. Small membrane vesicles are transported up the process by motor proteins, usually kinesins, which use microtubules as tracks as shown in Fig. 4.5. Upon reaching the ends of the central microtubule core, they detach. As mentioned previously, the membrane vesicles then bind to the myosin ATPase, and move along actin filaments from the cytosol to a position just underlying the plasma membrane. They are then inserted into the plasma membrane through SNARE-mediated fusion, which will be discussed later in this chapter.

One example of an axonal protein that is synthesized in the axonal growth cone is a specific member of the actin family, β-actin. This protein is synthesized in response to an attractive extracellular signal such as a tropic factor. Protein synthesis in axons and dendrites will be discussed in Chapter 5, Neuronal cell biology. The newly synthesized β-actin filament and other molecules form a motile network that can rapidly steer the growth cone in a different direction by a variety of mechanisms in response to changes in the concentration of intracellular signaling molecules.

One second messenger molecule directing by the growth cone is the concentration of cytosolic Ca^{++}. If a particular receptor differentially located on one surface of the growth cone encounters a gradient of its ligand, the cytoplasmic

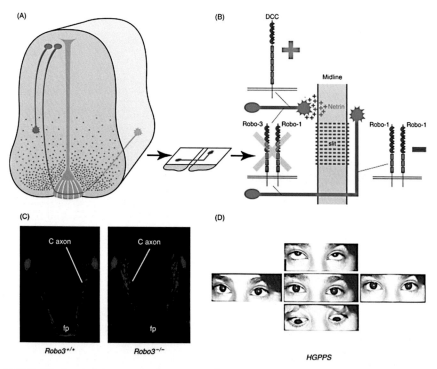

FIGURE 4.7 A switch from attraction to repulsion allows commissural axons to enter then leave the CNS midline.

(A) Schematic of a cross section through the mammalian spinal cord, illustrating the trajectory of commissural axons from the dorsal spinal cord to, and across, the ventral midline. The axons cross the midline at the floor plate; upon exiting the floor plate, they make a sharp right-angle turn and project in a rostral (anterior) direction. A long-range gradient of Netrin-1 protein (*green dots*) and a shorter-range gradient of Slit protein (*red dots*) are shown. Right: Illustration of the "open-book" configuration, in which the spinal cord is opened at the dorsal midline and flattened, which helps to visualize the crossing and turning behavior of the axons at the midline. (B) As the axons project to the midline ("precrossing"), they are attracted by Netrin-1 acting on the attractive Netrin receptor DCC. They are insensitive to the repulsive action of Slit proteins despite expression of the repulsive Slit receptors (Robo1 and Robo2) because they also express Robo3, a Slit receptor that suppresses the activity of Robo1 and Robo2. This enables these axons to enter the midline upon reaching it. However, upon crossing the midline, commissural axons downregulate expression of Robo3 and increase expression of Robo1 and Robo2 (not shown), which allows them to sense the repulsive action of Slits and results in their expulsion from the midline. (C) In Robo3 mutant mice, the precrossing commissural axons are prematurely sensitive to Slits, which prevents them from entering the midline. Shown are cross sections through a normal E11.5 mouse embryo (left) and a stage-matched Robo3 mutant embryo (right), illustrating the inability of commissural axons in the mutant to cross the midline. (D) Human patients with horizontal gaze palsy and progressive scoliosis (HGPPS) carry mutations in the human *ROBO3* gene. They are capable of coordinated eye movements along the vertical axis (center), but not along the horizontal axis (left and right panels are attempts by this HGPPS patient to look left and right). The deficits in horizontal gaze result from a defect in neuronal connections across the midline in the hindbrain, which is presumed to arise from aberrant axon guidance during development, since a similar defect is observed in Robo3 mutant mice.

Squire LR, Berg D, Bloom FB, Du Lac S, Ghosh A, Spitzer NC. Fundamental Neuroscience. 4th ed. Oxford: Elsevier; 2013 [Chapters 6–9, 16]. Fig. 16.11, p. 377. With permission.

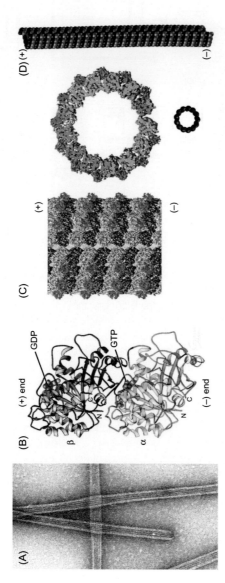

FIGURE 4.8 Structure of the α-tubulin/β-tubulin dimer and the microtubule.

(A) Electron micrograph of negatively stained microtubules. (B) Ribbon diagram with space filling guanosine triphosphate (GTP) on α-tubulin and guanosine diphosphate (GDP) on β-tubulin. Each subunit consists of approximately 450 residues arranged in two domains. Each domain is a β-sheet flanked by α-helices. The nucleotides bind in pockets similar to the binding site for nicotinamide adenine dinucleotide (NAD) on the enzyme glyceraldehyde-3-phosphate dehydrogenase. (C) Reconstructions from electron micrographs of frozen GTPγS (a slowly hydrolyzed analog of GTP) microtubules decorated with the small protein EB3 (*blue*) showing a side view and cross section. (D) Drawing of the microtubule used throughout this book. Both (C) and (D) show the longitudinal seam between two protofilaments, which breaks the helical repeat of tubulin dimers.

Pollard TD, Earnshaw WC, Lippincott-Schwartz J, Johnson GT. Cell Biology. 3rd ed. Philadelphia, PA: Elsevier; 2017 [Chapters 1, 3, 13–16, 21–24, 27, 30, 33, and 34]. Fig. 34.4, p. 596.

Ca^{++} on that side of the growth cone rises. The rise in cytoplasmic Ca^{++} in turn triggers growth of filopodia and lamellipodia through synthesis and insertion of actin filaments on that side, combined with new membrane insertion. This in turn produces a change in direction of the process. In contrast if differentially localized receptors for a repulsive ligand are activated, they cause a decreased Ca^{++} concentration on that side. This in turn triggers the turning away of the process. Thus the growth cone is constantly adjusting its direction of growth until it reaches its ultimate target.

Besides Ca^{++} there are many other intracellular messengers active in the growth cone. Among the best studied are cyclic AMP and cyclic GMP, which in many situations function in a yin—yang manner. Ultimately all of these second messenger molecules lead to changes in the cytoskeleton, which then translate into a changed direction of growth. Once the growing process reaches its target, a mature synapse is created.

4.2 THE SYNAPSE

Many presynaptic endings contain numerous small clear vesicles, called synaptic vesicles. These are filled with neurotransmitters: Glu, the excitatory transmitter, or GABA, the inhibitory transmitter (Fig. 4.9). The Glu vesicles are spherical while the GABA vesicles are oval; this allows them to be distinguished from one another in electron micrographs. Also, some presynaptic endings contain larger electron-dense vesicles filled with either small neurotransmitters including dopamine, norepinephrine, or serotonin, or with small neuropeptides. Many GABAergic synapses also contain one or more dense core vesicles.

One neuropeptide found in many GABAergic presynaptic endings is calretinin. Others include substance P, neuropeptide Y, FMRF amide, and opioids including enkephalins and dynorphins. Hormones that are found in other parts of the body also function as neuropeptides in the brain. These include insulin, IGF 1 and 2, secretin, gastrin, and somatostatin.

Each presynaptic ending contains many synaptic vesicles. Some are located directly underneath a specialization of the plasma membrane called the active zone. The active zone contains two parallel rows of large intramembranous particles consisting of voltage-activated Ca^{++} channels (Fig. 4.10).

Each synaptic vesicle contains one or more copies of a number of integral membrane proteins including VAMP 1/synaptobrevin and a Ca^{++} binding protein, synaptotagmin 1 (Fig. 4.11). These proteins together with two presynaptic integral membrane proteins, syntaxin 1 and SNAP 25, form the SNARE complex. This complex in turn triggers membrane fusion leading to neurotransmitter release.

The two neuron-specific integral plasma membrane SNARES, syntaxin 1 and SNAP-25, are found near the active zone (Fig. 4.12). The synaptic vesicles

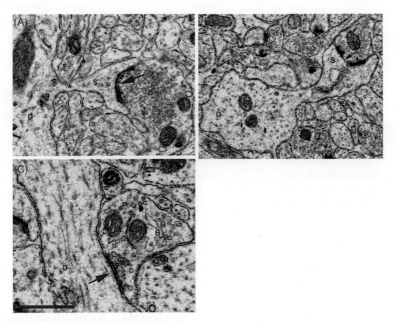

FIGURE 4.9 Ultrastructure of dendritic spines and synapses in the human brain.

(A and B) Narrow spine necks (*asterisks*) emanate from the main dendritic shaft (D). The spine heads (S) contain filamentous material (A and B). Some large spines contain cisterns of a spine apparatus (B). Asymmetric excitatory synapses are characterized by thickened postsynaptic densities (*arrows* A and B). A perforated synapse has an electron-lucent region amidst the postsynaptic density (*small arrow*, B). The presynaptic axonal boutons (B) of excitatory synapses usually contain round synaptic vesicles. Symmetric inhibitory synapses (*arrow*, C) typically occur on the dendritic shaft (D) and their presynaptic boutons contain smaller round or ovoid vesicles. Dendrites and axons contain numerous mitochondria (m). Scale bar = 1 μm (A and B) and 0.6 μm (C).

Squire LR, Berg D, Bloom FB, Du Lac S, Ghosh A, Spitzer NC. Fundamental Neuroscience. *4th ed. Oxford: Elsevier; 2013 [Chapters 6–9, 16]. Fig. 3.3, p. 44. With permission.*

clustered near the active zone are in close contact with these proteins (Fig. 9.4). There are also many other synaptic vesicles in the presynaptic ending, making up the reserve pool that can be recruited to the active zone when needed (see Fig. 9.1).

As described previously, synaptic vesicles are anchored to the synaptic plasma membrane at the active zone. These anchorage sites are held together by the three SNARES. Because of their specificity and their tight binding to each other, the SNAREs can be thought of as receptors. The three proteins interact to form a parallel four-stranded α-helix that brings the two membranes very close together (Fig. 4.12). The binding of the three SNARES to each other occurs spontaneously, requiring no energy. This is because the four helix-forming surfaces of the

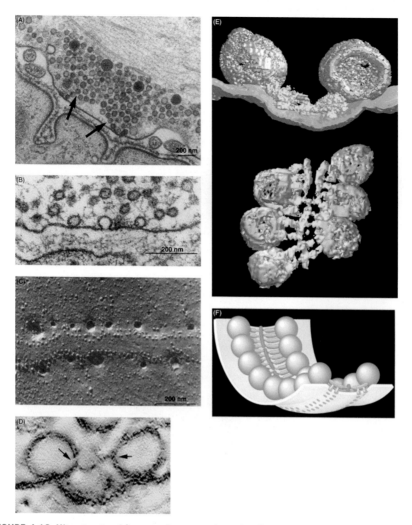

FIGURE 4.10 Ultrastructural images of exocytosis and active zones.

(A–C) Synapses from frog sartorius neuromuscular junctions were quick-frozen milliseconds after stimulation in conditions that enhance transmission. (A) A thin section showing vesicles clustered in the active zone (*arrows*), some docked at the membrane. (B) Shortly (5 ms) after stimulation, vesicles were seen to fuse with the plasma membrane. (C) After freezing, presynaptic membranes were freeze-fractured and a platinum replica was made of the external face of the cytoplasmic membrane leaflet. Vesicles fuse about 50 nm from rows of intramembranous particles thought to include Ca^{++} channels. (D–F) The fine structure of the active zone at a frog neuromuscular junction as seen with electron tomography. (D) In a cross-sectional image from tomographic data, two vesicles are docked at the plasma membrane and additional electron-dense elements are seen. When these structures are traced and reconstructed through the volume of the EM section (E), proteins of the active zone (*gold*) appear to form a regular structure adjacent to the membrane that connects the synaptic vesicles (*silver*) and plasma membrane (*white*). Viewed from the cytoplasmic side (E, lower image), proteins are seen to extend from the vesicles and connect in the center. (F) Schematic rendering of an active zone based on tomographic analysis. An ordered structure aligns the vesicles and connects them to the plasma membrane and to one another.

Squire LR, Berg D, Bloom FB, Du Lac S, Ghosh A, Spitzer NC. Fundamental Neuroscience. *4th ed. Oxford: Elsevier; 2013 [Chapters 6–9, 16]. Fig. 7.1, p. 141. With permission.*

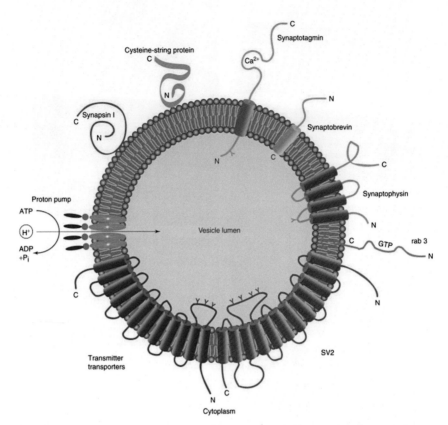

FIGURE 4.11 Schematic representation of the structure and topology of major synaptic vesicle membrane proteins.

Squire LR, Berg D, Bloom FB, Du Lac S, Ghosh A, Spitzer NC. Fundamental Neuroscience. 4th ed. Oxford: Elsevier; 2013 [Chapters 6–9, 16]. Fig. 7.4, p. 148. With permission.

SNARES contain many hydrophobic amino acids that prefer to interact with one another than to be exposed to a hydrophilic environment.

The synaptic vesicle Ca^{++} binding protein synaptotagmin 1 serves as a trigger preventing the synaptic vesicles from fusing with the presynaptic membranes until the Ca^{++} rises to a near-millimolar level in the cytoplasm directly underlying the active zone. Now the two membranes can fuse and neurotransmitter release occurs (Fig. 4.12). The two scientists who described the SNARE mechanism, James Rothman and Randy Scheckman, together with Thomas Sudhoff, who worked out the mechanism by which synaptophysin activates membrane fusion, were collectively awarded the Nobel Prize in Physiology and Medicine in 2013.

The synaptic vesicle fusion with the synaptic plasma membrane occurs when an electric impulse reaches the axonal ending, causing the membrane potential to be depolarized. This depolarization causes voltage-sensitive Na^+ channels to

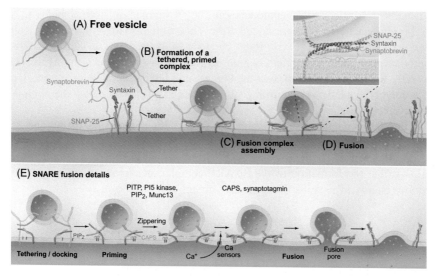

FIGURE 4.12 Membrane fusion by SNARE proteins.

The example is neurotransmitter release. (A) A synaptic vesicle with tethers and v-SNARE synaptobrevin (*blue*) approaches the plasma membrane, which has tethers and complexes of the t-SNAREs syntaxin (*red*) and SNAP-25 (*green*). (B) The v- and t-SNAREs dock to form a primed but inactive complex. (C) When stimulated by an action potential, the SNAREs zip together to form a trans-SNARE complex that pulls the membranes together. The inset in the upper right shows a ribbon diagram of the fully assembled SNAREs anchored to the membranes. (D) The vesicle fuses with the plasma membrane, leaving cis-SNARE complexes on the plasma membrane. (E) Overview of SNARE-mediated fusion. SNARE, soluble *N*-ethylmaleimide-sensitive factor attachment protein receptor.

Pollard TD, Earnshaw WC, Lippincott-Schwartz J, Johnson GT. Cell Biology. 3rd ed. Philadelphia, PA: Elsevier; 2017 [Chapters 1, 3, 13–16, 21–24, 27, 30, 33, and 34]. Fig. 21.14, p. 364. With permission.

open. This in turn produces the opening of the selective Ca^{++} channels in the active zone. The type of exocytosis in which membrane fusion occurs only after the cytoplasmic Ca^{++} concentration is significantly raised is called regulated exocytosis, as was described in Chapter 1, Classes of receptors, their signaling pathways, and their synthesis and transport. Although regulated exocytosis occurs in many different cell types, it has been studied at its most fundamental molecular level in neurons.

As discussed in Chapter 1, Classes of receptors, their signaling pathways, and their synthesis and transport, a specific member of the Rab family, a small GTPase, is necessary for vesicles to undergo fusion with the correct plasma membrane. In the case of the neuronal synaptic vesicles the Rab required is Rab 3.

When the synaptic vesicle approaches its target membrane, Rab 3 binds to an active zone anchored receptor called Rim. This protein, part of a tethering complex, specifically binds to Rab 3 and activates its GTPase activity. Simultaneously with membrane fusion the Rab 3—associated GTP is hydrolyzed to GDP, Rab 3 leaves the plasma membrane, and after several steps, reassociates with another synaptic vesicle to repeat the process. Fig. 4.13 shows the complete Rab cycle.

After the synaptic vesicle membrane fuses with the plasma membrane, the SNARE complex separates through the action of *N*-acetyl maleamide sensitive factor, an ATPase. The reason that ATP hydrolysis is essential for complex separation is that the formation of the SNARE complex occurs spontaneously, and it requires energy to separate the respective complex.

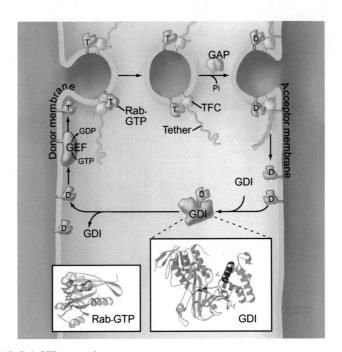

FIGURE 4.13 Rab GTPase cycle.

Rab-GDP is bound to GDI in the cytoplasm. A GTPase dissociation factor (not shown) on the membrane separates Rab-GDP from GDI so it can be activated by a membrane-associated, Rab-specific GEF. Rab-GTP recruits effectors, such as coiled-coil tethers and TFCs, which aid in targeting and docking the vesicle. A GAP stimulates GTP hydrolysis and returns Rab-GDP to GDI in the cytoplasm. Insets show ribbon diagrams of Rab-GTP and GDI. *GDI*, guanine nucleotide dissociation inhibitor; *GDP*, guanosine diphosphate; *GTP*, guanosine triphosphate; *GTPase*, guanosine triphosphatase; *Pi*, inorganic phosphate; *TFC*, tethering factor complex.

Pollard TD, Earnshaw WC, Lippincott-Schwartz J, Johnson GT. Cell Biology. *3rd ed. Philadelphia, PA: Elsevier; 2017 [Chapters 1, 3, 13–16, 21–24, 27, 30, 33, and 34]. Fig. 21.6, p. 357. With permission.*

The synaptic vesicle-containing membrane now moves away from the active zone and is internalized in a large uncoated vesicle together with other synaptic vesicle membranes. They are then coated with the AP2 adapter and clathrin to form clathrin-coated vesicles. Then the two coat proteins are removed (Fig. 4.14). After uncoating, the vesicles are refilled with their specific neurotransmitter through the action of a specific vesicular pump. For example, a dopaminergic synaptic vesicle contains integral membrane proteins that pump dopamine into the acidified lumen using ATP hydrolysis to provide the necessary energy (Fig. 4.15).

Synaptic vesicle recycling is necessary because in long projection nerves it requires days or weeks for a newly synthesized synaptic vesicle to reach the pre-synaptic terminal. Without recycling the synaptic vesicles would be rapidly depleted and neurotransmitter release would cease. It is estimated that each synaptic vesicle is recycled about 1000 times before it is degraded.

Once the neurotransmitters are released, with some percentage binding to their receptors on the postsynaptic membrane, a number of different actions result for the remainder, depending on the neurotransmitter. A portion of the released glutamate is taken up by astrocytes that surround the synaptic cleft as will be described in Chapter 6, Glial cells; the evolution of the myelinated axon; blood–brain barrier. Some of the released GABA is taken back into the presynaptic terminal by the action of a GABA transporter. Acetylcholine (ACh) is broken down rapidly by acetylcholinesterase. Then the choline is taken up by

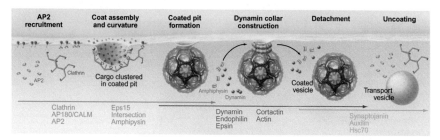

FIGURE 4.14 Timeline of receptor-mediated endocytosis driven by the clathrin-coated vesicle.

AP2 complexes are targeted to the plasma membrane and interact with tyrosine-based sorting motifs in the cytoplasmic domains of transmembrane proteins. AP2 also initiates the assembly of a polygonal lattice of clathrin, which concentrates cargo molecules in coated pits. AP2 and clathrin interact with the BAR-domain protein amphiphysin, which attracts the guanosine triphosphatase (GTPase) dynamin. After assembly of a coated pit and fission of the coated vesicle from the plasma membrane, auxilin and the adenosine triphosphatase (ATPase) Hsc70 dissociate clathrin and release the vesicle carrying cargo into the cell for fusion with endosomes.

Pollard TD, Earnshaw WC, Lippincott-Schwartz J, Johnson GT. Cell Biology. 3rd ed. Philadelphia, PA: Elsevier; 2017 [Chapters 1, 3, 13–16, 21–24, 27, 30, 33, and 34]. Fig. 22.9, p. 384. With permission.

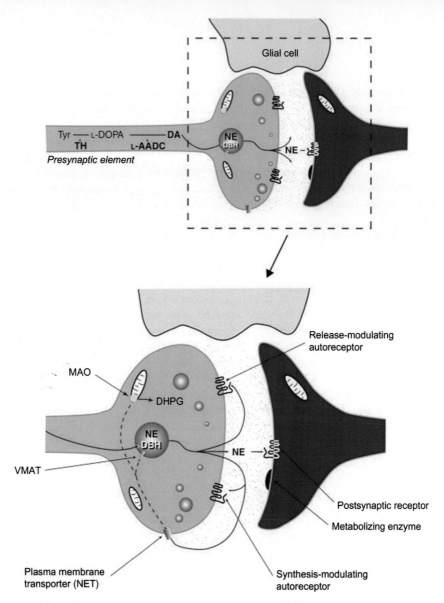

FIGURE 4.15 Characteristics of a norepinephrine (NE)-containing catecholamine neuron.

Tyrosine (Tyr) is accumulated by the neuron and then is metabolized sequentially by tyrosine hydroxylase (TH) and L-aromatic amino acid decarboxylase (L-AADC) to dopamine (DA). The DA is then taken up through the vesicular monoamine transporter into vesicles. In DA neurons, this is the final step. However, in this NE-containing cell, DA is metabolized to NE by dopamine-*b*-hydroxylase (DBH), which is found in the vesicle. Once NE is released, it can interact with postsynaptic noradrenergic receptors or presynaptic noradrenergic autoreceptors. The accumulation of NE by the high-affinity membrane NE transporter (NET) terminates the actions of NE. Once taken back up by the neuron, NE can be metabolized to inactive compounds (DHPG) by degradative enzymes such as monoamine oxidase (MAO) or taken back up by the vesicle.

Squire LR, Berg D, Bloom FB, Du Lac S, Ghosh A, Spitzer NC. Fundamental Neuroscience. 4th ed. Oxford: Elsevier; 2013 [Chapters 6–9, 16]. Fig. 6.2, p. 121. With permission.

the choline transporter into the presynaptic terminal and ACh is resynthesized from acetate and choline by ACh synthase. The monamines dopamine, norepinephrine, and serotonin are returned to their respective presynaptic terminal by either the dopamine, epinephrine, or serotonin transporter. Once the transmitters have been taken back into their respective terminals and if required, resynthesized, they are pumped back into their synaptic vesicle by the action of vesicular transporters specific for each respective neurotransmitter as described previously.

Neuropeptides once released cannot be reutilized at the synapse because there is no reuptake mechanism. They also signal via G-protein coupled receptors (GPCRs) and G proteins and these signaling events are much slower than are transmitters that act by binding to ionic receptors. For these reasons neuropeptides serve a modulatory role in neurotransmission in more evolved organisms, rather than acting as primary transmitters as they do in smaller, less evolved organisms like Hydra. Because of their small size, these organisms do not require instantaneous transfer of information from one neuron to another.

There are two key adhesion proteins that connect the pre- and postsynaptic terminals. One is the integral membrane protein Neurexin, on the presynaptic membrane, the other is the adhesion protein Neuroligin, on the postsynaptic membrane. They serve as receptors for each other and help to coordinate and develop the pre- and postsynaptic elements. Beside Neurexin and Neuroligin, other connecting proteins are present that are very sensitive and selective. They also can be regarded as receptors that recognize surface ligands on the adjacent cell membrane. These proteins include members of the Cadherin and Protocadherin families. These two adherence protein families will be described in detail in Chapter 12, Key receptors involved in laminar and terminal specification and synapse construction.

As described previously, the synapse is composed of a presynaptic element at the end of the nerve axon, and a postsynaptic element at the end or on the sides of a dendrite. These structures are separated by a synaptic cleft containing a specialized type of ECM component. There can also be axo-axonal synapses but they are much less common.

On either side of the cleft there are connecting proteins anchoring the pre- and postsynaptic elements to one another. Astrocytes, the most abundant glial cell, which will be described in detail in Chapter 6, Glial cells; the evolution of the myelinated axon; blood—brain barrier, also help to seal off the synapse from other portions of the extracellular space.

An important element of the postsynaptic region directly opposite the presynaptic ending is the postsynaptic density. It contains a signaling enzyme called calmodulin kinase II, which is the most abundant protein in the postsynaptic density. This kinase phosphorylates a number of important proteins when the Ca^{++} concentration in the postsynaptic cytosol rises. This increase is caused by the influx of Ca^{++} through NMDA channels, or through the action of metabotropic GPCRs including Glu receptors that are anchored to the periphery of the synapse (Fig. 4.16). Many postsynaptic

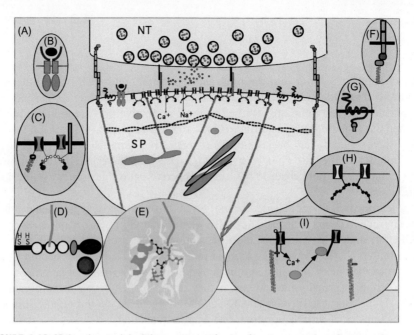

FIGURE 4.16 Molecular model of the postsynaptic density at a central excitatory synapse.

(A) Hypothetical organization of presynaptic (NT, nerve terminal) and postsynaptic (SP, spine) structures. Synaptic vesicles (*orange spheres*) release glutamate into the synaptic cleft, which in turn stimulates NMDA (*blue rectangle*), AMPA (*red, yellow rectangle*), and metabotropic (*brown membrane protein*) glutamate receptors. In the spine, actin cables (*vertical pink filaments*) are linked to brain spectrin (*red, horizontal molecules*). Also present in the spine are endoplasmic reticulum (*blue membranous structure*) and calmodulin (*green ovals*). CAM kinase II, other kinases, and proteases are present. (B) TrkB responds to BDNF (*pink receptor; blue ligand*). (C) The neuroligin (*green rectangle*) cytoplasmic domain binds PSD-95 (*blue, green, yellow ovals*). PSD-95 binds GKAP/SAPAP/DAP protein (*red*) and the NMDA receptor (*blue rectangle*), which in turn binds α-actinin (*orange*) and actin (*pink*). (D) The MAGUK protein PSD-95 contains three N-terminal PDZ domains (*yellow circles*), an SH3 domain (*green oval*), and a guanylate kinase domain (*blue oval*), which binds members of the GKAP/SAPAP/DAP protein family (*red circle*). The N terminus contains a pair of cysteine residues (SH), which function in dimerization. (E) The crystal structure of a PDZ domain of PSD-95 (*yellow ribbon*) shows helix αB (*green segment of ribbon*) with specific residue αB1 (*blue histidine*) that hydrogen bonds to a threonine at position −2 (*red residue*) of the C terminus (*olive line and residues*) of the NMDA receptor NR2A and NR2B subunits. (F) Cadherins (*orange transmembrane protein*) are linked by their cytoplasmic domain (*blue*) to β-catenin (*red*) and α-catenin (*green*) and actin (*pink*). (G) Metabotropic GluR binds Homer (*blue*) via PDZ domain (*circle*). (H) Two AMPA receptors (*red, yellow rectangle*) are shown. Each binds GRIP, which has seven PDZ domains (*red circles*). Dimerization of GRIP via N termini is hypothetical. Ca^{++} that enters the spine through the NMDA receptor in response to receptor binding of glutamate (*yellow circle*) is proposed to activate calmodulin (*green oval*), which displaces α-actinin and actin (*orange oval, pink chain*) from the NMDA receptor NR1 subunit C terminus.

Modified from Ziff EB. Enlightening the postsynaptic density. Neuron. *1997;19:1163–74. Fig. 4. PMID: 9427241. With permission.*

densities lie just beneath the plasma membrane apposing the presynaptic active zones. A group of proteins called PDZ proteins (because they have one or more PDZ domains) bind to Glu or GABA receptors as well as to other PDZ containing proteins. They are the key organizing elements for the postsynaptic domains (Fig. 4.16).

A key adapter protein in excitatory synapses is PSD-95, which contains three PDZ domains that attach it to the TARP protein, a transmembrane protein, which is in turn linked to the AMPA receptor. Another PDZ domain on a second PSD-95 molecule is directly linked to the NMDA receptor (Fig. 4.16).

As mentioned above, the metabotropic Glu receptors are anchored to the periphery of the synapse. They bind to the adapter Homer, which in turn binds to Shank, attaching it to another adapter, GKAP, which in turn binds to PDZ-95.

A second class of synapse found in the brain on both neurons and astrocytes and other glial cells is the electrical synapse. Electrical synapses convey charges incredibly rapidly and are used in cases that require an extremely fast response. An example is the tail-flip response in goldfish, which is produced by a neuron called the Mauthner cell found in the brain stem. This cell is electrically coupled to many sensory neurons. Following almost instantaneous activation upon touch, the Mauthner cell signals to motor neurons in the tail, which in turn cause the fish to escape from danger. Also many cells can be electrically coupled allowing synchronous responses throughout a whole group of neurons and/or glia. The electrical signal is graded and therefore does not require an action potential.

Electrical synapses, also called gap junctions, occur at plasma membrane specializations in which the two membrane surfaces are only 3.5 nm apart instead of 20. At these specializations on each cell surface, hexagonal arrays of proteins called connexins form hemiconnexins. These hemiconnexins contact an identical array of hemiconnexins to form a connexon. Every hemiconnexin is composed of six identical subunits, each composed of four membrane-spanning domains (Fig. 4.17). These interactions are exceedingly sensitive and specific and can be compared to receptor−ligand interactions.

The connexons form pores with diameters of 1.5−2 nm. The pores are large enough for small ions including Na^+ and Ca^{++} as well as larger molecules including the intracellular second messengers cAMP and cGMP, to flow from one cell to another. These connexons also allow trophic and other behavioral modifying molecules to rapidly move from one cell to another. The connexons at rest are thought to have a closed pore in the center, which opens when an appropriate signal is received (Fig. 4.17).

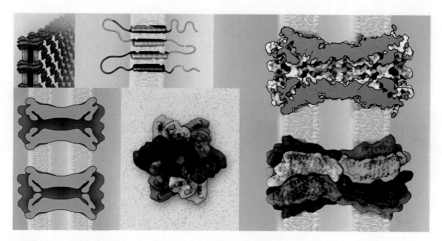

FIGURE 4.17 Structure of the gap junction connexon channel.

(A) Drawing of gap junction connexons forming channels between the cytoplasms of adjacent cells. (B) Transmembrane topology of connexins. Four α-helices cross the lipid bilayer. Conserved residues (*maroon*) form the transmembrane and extracellular loops are required for channel assembly. Cytoplasmic loops between helix 2 and helix 3 and the C-terminal tails vary in length among connexin isoforms. Removal of the C-terminal tail from connexin 43 alters its gating properties. (C) Diagram showing how the N-terminal α-helices of Cx26 may form a plug that blocks the pore of the closed channel. (D and E) Crystal structure of the connexin 26 the gap junction channel. (C and D) Top and side views of a ribbon diagram with each of the six subunits a different color. The two extracellular loops of each subunit associate with the other half channel to span the 4-nm gap between the membranes. The upper panel of (D) shows as a space filling model cut through the middle of the transmembrane pore.

Pollard TD, Earnshaw WC, Lippincott-Schwartz J, Johnson GT. Cell Biology. *3rd ed. Philadelphia, PA: Elsevier; 2017 [Chapters 1, 3, 13–16, 21–24, 27, 30, 33, and 34]. Fig. 31.7, p. 549. With permission.*

RECOMMENDED BOOK CHAPTERS

Kandel ER, Schwartz JH, Jessel TM, Siegelbaum SA, Hudspeth AJ. *Principles of Neural Science*. 5th ed. New York: McGraw Hill; 2013 [Chapters 8, 12, 54].

Squire LR, Berg D, Bloom FB, Du Lac S, Ghosh A, Spitzer NC. *Fundamental Neuroscience*. 4th ed. Oxford: Elsevier; 2013 [Chapters 6–9, 16].

Review Articles

Südhof TC. The molecular machinery of neurotransmitter release (Nobel lecture). *Angew Chem Int Ed Engl*. 2014;53:12696–12717.

Zheng JQ, Poo MM. Calcium signaling in neuronal motility. *Annu Rev Cell Dev Biol*. 2007;23:375–404.

Ziff EB. Enlightening the postsynaptic density. *Neuron*. 1997;19:1163–1174.

Neuronal cell biology

5

CHAPTER OUTLINE

As expected for the major information-carrying cell of the most complex organ, the neuron is the most complicated cell type in the body. Neurons have a myriad of sizes and shapes, some with many dendrites, some with only a few. Essentially all neurons, however, have only one major axon but it can have several branches (Fig. 5.1). Of the three subcompartments comprising the neuron, two, the axon and the dendrite, are unique and differ in many ways from each other as well as from the neuronal cell body.

The dendrite is unique in several important respects. It transports mRNAs for hundreds of abundant proteins to the postsynaptic endings. These included actin, a key component of the motile apparatus of cells, and calmodulin kinase II, the most abundant protein making up the postsynaptic density underneath the post-synaptic surface of dendrites. The mRNAs, after transport to the postsynaptic cytoplasm in a complex called ribonucleic protein particles, are stored on ribosomes located in the cytoplasm and ultimately translated into proteins. In the case of β-actin, a very abundant dendritic protein, a 54-nucleotide binding sequence contained in the $3'$ untranslated region of the β-actin mRNA serves to bind the ZIP1 protein, which functions as a "zip code." This complex is then specifically transported to dendrites. After reaching its destination, a tyrosine residue on ZIP1 is phosphorylated by the Src kinase, allowing the mRNA to be translated. Other dendritically transported mRNAs are also thought to contain specific "zip codes" but these remain to be discovered. The sensitivity and binding of the ZIP protein to the zip code containing mRNA can also be considered an interaction between a protein receptor and an RNA ligand.

Dendrites contain endoplasmic reticulum (ER) and Golgi elements, enabling the dendrite postsynaptic terminals to synthesize plasma membrane proteins. This allows each terminal to respond very rapidly to changes in its environment. It is known that dendritic terminals, also known as dendritic arbors, can grow, contract, or even disappear during development and in later life. This process is referred to as synaptic pruning during brain development, and synaptic plasticity after the brain matures. Synaptic pruning will be discussed in detail in Chapter 15, Synaptic pruning during development and throughout life depends on

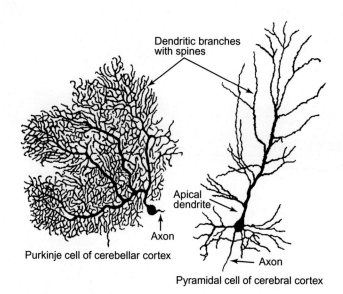

FIGURE 5.1 Typical morphology of projection neurons.

(Left) A Purkinje cell of the cerebellar cortex and (right) a pyramidal neuron of the neocortex. These neurons are highly polarized. Each has an extensively branched, spiny apical dendrite, shorter basal dendrites, and a single axon emerging from the basal pole of the cell. The scale bar represents approximately 200 μm.

Squire LR, Berg D, Bloom FB, Du Lac S, Ghosh A, Spitzer NC. Fundamental Neuroscience. *4th ed. Oxford: Elsevier; 2013 [Chapters 4, 5]. Fig. 3.1, p. 42. With permission.*

trophic factor interactions: the corollary described in Chapter 14, Postulating the quantization of trophic factor requirements, provides a plausible explanation for the survival of neurons partially deprived of their full complement of trophic factors.

Dendrites are unique in two other respects. They have microtubules, a key organelle that together with the motor proteins, kinesin and dynein, move proteins and membrane vesicles and mitochondria from one part of the dendrite to another. Other cells in the body have microtubules oriented so that all of their minus ends are near the cell nucleus and their plus ends are near the plasma membrane. In contrast, dendrites contain microtubules having polarity in both directions, plus to minus and minus to plus, with respect to the plasma membrane. Finally, only dendrites contain the microtubule cross-linking protein, MAP2, but not the other major microtubule cross-linking protein, tau, which is found only in axons (Fig. 5.2). This fact allows the easy identification of axons and dendrites by immune staining with specific antibodies against the two proteins.

Axons have several unique features as well. Because axons as opposed to dendrites can be 50 ft. long in blue whales' motor neurons, they have evolved a system called rapid axonal transport to convey membranes and their cargoes from

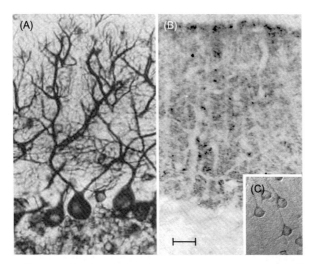

FIGURE 5.2 Distributions of Tau and MAP 2 in sections of the cerebellum from rat brain labeled with antibodies and subjected to histochemical staining.

(A) MAP2 (*black*) is concentrated in the cell bodies and dendrites of Purkinje cells. (B) Tau is concentrated in the axons of granule cells, which appear here as small dark dots. (C) Tau staining in the cell body and neurites of a pyramidal cell from another part of the brain. Scale bar is 20 μm.

Pollard TD, Earnshaw WC, Lippincott-Schwartz J, Johnson GT. Cell Biology. 3rd ed. Philadelphia, PA: Elsevier; 2017 [Chapters 1, 3, 13–16, 21–24, 27, 30, 33, and 34]. Fig. 34.11, p. 602. With permission.

the cell body to the presynaptic ending and back. The rate of movement of these vesicles along microtubules propelled by the ATPase kinesin toward the presynaptic ending is 4 cm/day, which is at least 100 times faster than similar movements in other cells. This rapid movement occurs even though the tracks, microtubules, and the locomotives, kinesin or dynein, are the same in all cells. While no one understands the molecular basis of rapid axonal transport, an analogous situation may be a comparison of Japanese bullet trains with Amtrak trains. Both employ similar tracks and engines but in the former case the tracks are straighter and the stops much less frequent.

The membrane associated vesicular proteins, which move to the axon terminal by rapid axonal transport, can be divided into at least five major classes. The first is synaptic vesicles, which carry their transmitters to the axon terminals. The second class is composed of membrane molecules, which comprise the presynaptic terminals. These include Ca^{++} channels, Na^+ channels, and transporters that pump released neurotransmitters including dopamine, norepinephrine, and serotonin back into the presynaptic terminal. These transmitters can then be reutilized after being pumped into their respective transmitter-containing synaptic vesicle by a separate class of transporter, which is an integral component of the synaptic vesicle (see Fig. 4.15). The third class consists of dense core vesicles that carry dopamine, serotonin, norepinephrine, or one of many neuropeptides. The fourth class consists of mitochondria, the energy producing organelles, at least one of which is found at each presynaptic ending. Finally there appears to be a vesicle class that carries membrane proteins to specific regions along the axon. In myelinated axons these sites must be at nodal regions along the axon because they are the only regions where membrane fusion can occur. Three proteins in this class are Na^+ and K^+ channels and Na^+, K^+ ATPases that maintain the resting potential of the axon. The membrane proteins of the myelinated projection axons will be discussed in Chapter 6, Glial cells; the evolution of the myelinated axon; blood−brain barrier.

There is evidence that the amyloid precursor protein involved in Alzheimer's disease links some vesicles bound for the axonal plasma membrane to kinesin light chains. The AMPA Glu receptor is carried in vesicles linked through the GRIP1 complex to kinesin heavy chains. Some synaptic vesicle precursors bind kinesin through syntabulin and syntaxin 1. Also the GABAA receptor−carrying vesicles bind to the HAP-1/kinesin complex. This binding is disrupted by mutant Huntingtin, the protein that causes Huntington's disease. All of these specific interactions between proteins on the vesicle surface and a motor molecule can be considered comparable to receptor−ligand interactions.

Retrograde neurotrophin-Trk containing vesicles associate with dynein in the cytoplasm of postsynaptic endings via Rab 6. These vesicles then are retrogradely transported to the neuron nucleus where they turn on genes supplying trophic support for the neuron. The speed of this transport is 22 cm/day, more than five times faster than the rate of anterograde rapid transport. To give us an idea of how fast these transport rates are, if a vesicle were a small automobile, it would travel at 50 miles/h in the anterograde direction or 220 miles/h in the retrograde direction.

Beside membrane proteins moving by rapid transport, most cytoskeletal and cytosolic proteins travel down the axon and dendrites by a much slower mechanism called slow transport. The speed of slow transport is approximately 50 times slower than the speed of rapid transport. It was first believed that the slow transported proteins move by simple diffusion; but in recent years with the development of much more sensitive imaging techniques, it is clear that at least some slow transported components actually move on kinesin and dynein motors attached to microtubules. The slow components seem to make many more stops, detachments from microtubules, and even reversals than do the fast transported components.

There is also now strong evidence that there is the synthesis of hundreds of distinct proteins in axons at the presynaptic endings and at the nodes of Ranvier in myelinated axons as well. One major reason that axons employ local protein synthesis is that due to their length, they cannot synthesize a given protein in the cell body, transport it a long distance to the axon ending, and replace an identical protein that has been destroyed, in the time required to keep the terminal functional. Even at the speed of 4 cm/day, the rate of anterograde rapid transport, it can take several weeks to transport a vesicle to the end of a long axon such as a motor neuron that innervates the human toe. A second major reason is that the axon, especially its terminals, needs to constantly modify the plasma membrane, as well as the contractile and signaling apparatus to accommodate changes in the environment. A good example is the synaptic response to rapidly repeating membrane depolarizations, causing an increased firing rate, and therefore the need for more synaptic vesicles.

Transporting mRNAs in the form of a ribonucleic particle and then storing them until needed appears to be a very effective strategy for the functions that these axons and presynaptic endings carry out. The presynaptic endings contain all the machinery to translate thousands of different mRNAs into proteins. These include ribosomes, tRNAs, ER, and Golgi apparatus. However, the vast majority of these mRNAs are inactive at any one time, being bound to particles containing RNA and protein. These particles block their translation until the proteins they code for are needed. One example of an axonal protein that is synthesized in the axonal growth cone is β-actin, which is synthesized in response to an attractive extracellular signal such as a tropic factor.

Based on evidence gathered from the study of neuron precursors on cell culture dishes, neurons initially put out several processes. One is somehow selected to become an axon and the others become dendrites. The selection is based at least in part on the interaction of receptors on the processes with extracellular matrix components called proteoglycans and glycoproteins. These components play many key roles in development, which will be discussed in later chapters.

Once one neuronal process is determined to become the sole axon, a major sorting problem arises. It is known that the axonal and somatodendritic proteins differ greatly with regard to the plasma membrane and internal organelle components. How is that separation achieved?

Over the last decade, evidence has accumulated that two small areas adjacent to the axon hillock, the preaxonal exclusion zone (PAEZ) nearest the beginning of the axon, and the axon initial segment (AIS), are the key areas involved in this process.

The key to the selective sorting of the axonal and dendritic organelles depends on their differential affinities for microtubule-based motors at the PAEZ. One class of motor is comprised of members of the kinesin family. The vast majority of these proteins have their motor ATPase associated with the N-terminal segment of their heavy chains, and move in a minus to plus direction along microtubules that are oriented with their minus end in the cell body and their plus ends near the axon termini. The microtubule binding protein CAMSAP2 binds to the minus ends of microtubules throughout the neuron, stabilizing them. This protein is enriched in the PAEZ. Another protein, TRIM46, is critical to the organization of parallel bundles of microtubules extending from the PAEZ to the AIS. During neuronal development, TRIM46 localizes to the one neurite destined to become the axon.

The other class of microtubule-based motors, dynein, moves in a plus to minus direction. As stated above, the axonal microtubules all are oriented in a minus to plus direction. In contrast, the dendritic microtubules are distributed such that about 50% are minus to plus and the others plus to minus.

Thus when a newly synthesized organelle destined for the axon enters the PAEZ it binds to kinesin and moves along microtubules to its ultimate axonal destination. In contrast, if a newly synthesized somatodendritically destined organelle enters the PAEZ, it binds to a dynein molecule and moves backward into the cell body and in some cases into the dendritic compartment. If any organelles meant for the somatodendritic compartment escape, they are retrieved in the AIS and returned to their appropriate destination.

It appears that plasma membrane proteins and lipids destined for the axon are sorted from somatodendritic ones in the AIS. A ring of actin filaments is present here, which plays a role in this sorting process, although the molecular mechanism is still not well understood.

GENERAL REFERENCES

Recommended Book Chapters
Kandel ER, Schwartz JH, Jessel TM, Siegelbaum SA, Hudspeth AJ. *Principles of Neural Science*. 5th ed. New York, NY: McGraw Hill; 2013 [Chapters 3, 4].
Squire LR, Berg D, Bloom FB, Du Lac S, Ghosh A, Spitzer NC. *Fundamental Neuroscience*. 4th ed. Oxford: Elsevier; 2013 [Chapters 4, 5].

Review Articles
Baas PW, Lin S. Hooks and comets: The story of microtubule polarity orientation in the neuron. *Dev Neurobiol.*. 2011;71:403−418.
Holt CE, Schuman EM. The central dogma decentralized: new perspectives on RNA function and local translation in neurons. *Neuron*. 2013;80:648−657.

Glial cells, the myelinated axon, and the blood-brain barrier

CHAPTER OUTLINE

6.1 GLIAL CELLS

In addition to neurons, the other major class of cell in the brain is the glial cell. There are three types of glial cells in the central nervous system (CNS): astrocytes and oligodendrocytes, which together are referred to as macroglia, and microglia. In the peripheral nervous system (PNS), Schwann cells carry out a similar function to that of oligodendrocytes.

There is no evidence for glial cells in early animal species including Hydra and other Cnidarians. They are, however, found in species considered representative of both early vertebrates and invertebrates; though it is not clear whether the two classes of macroglia evolved from the same ancestor or from separate ones.

In mammalian brains there are approximately as many total glial cells as there are neurons. Glial cells, mainly astrocytes, develop after neurons have formed, and completely fill the gaps between them, with the exception of capillary endothelial cells (CECs), which are discussed below. Recent work has demonstrated that the number of glial cells in a given region of the brain is determined by the number needed to completely cover the surface of the earlier developing neurons. In mammals, larger species have larger neurons. Therefore because glial cells remain almost uniform in size throughout all mammals, larger mammals require more glia to completely cover the neuronal surfaces.

The type of glial cell that is the most heterogeneous, the astrocyte, is also the most versatile. At synapses, astrocytes surround the synaptic endings and help regulate the extracellular concentration of Na^+, K^+, Ca^{++}, Cl^-, and other ions necessary for homeostasis. To accomplish this task, they employ numerous transporters. Astrocytes are the only cell type in the brain that contains the AQP4 water channel. This channel is essential to maintain water homeostasis in the brain. Astrocytes also have receptors for neurotransmitters and can synthesize and

Receptors in the Evolution and Development of the Brain. DOI: https://doi.org/10.1016/B978-0-12-811012-6.00006-6

release many neurotransmitters and trophic substances through regulated vesicular exocytosis. These molecules can influence the development and functioning of neurons.

Another key function of astrocytes is to use surface glutamate transporters to take up glutamate that is released at the presynaptic endings of excitatory neurons. This prevents excess glutamate accumulation at the synapse, which can overstimulate the postsynaptic neuron leading to a form of cell death called excitotoxic cell death. After glutamate uptake, the astrocyte converts it to glutamine through the action of an enzyme called glutamine synthetase that is uniquely found in astrocytes. The glutamine is then released, taken up by neurons, and used to resynthesize glutamate (Fig. 6.1). Astrocytes also take up GABA released at presynaptic terminals and synthesize glutamine from it through the sequential action of glutamate synthase and glutamine synthase (Fig. 6.1).

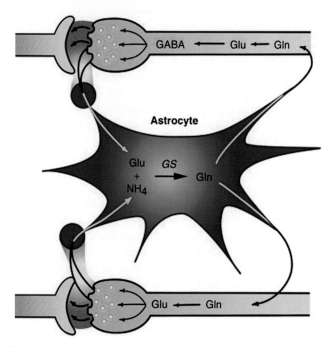

FIGURE 6.1 The glutamate—glutamine cycle is an example of a complex mechanism that involves an active coupling of neurotransmitter metabolism between neurons and astrocytes.

The systems of exchange of glutamine, glutamate, γ amino butyric acid (GABA), and ammonia between neurons and astrocytes are highly integrated. The postulated detoxification of ammonia and the inactivation of glutamate and GABA by astrocytes are consistent with the exclusive localization of glutamine synthetase in the astroglial compartment.

Squire LR, Berg D, Bloom FB, Du Lac S, Ghosh A, Spitzer NC. Fundamental Neuroscience. 4th ed. Oxford, UK: Elsevier; 2013 [chapters 3, 4, 12], Fig. 3.14, p. 54. With permission.

Astrocytes produce lactate from pyruvate. During periods of high neurotransmitter release, the astrocyte-released lactate is used as the surrounding neurons' chief energy source. Therefore it is essential that lactate is released during memory formation and other neuronal functions requiring constant neurotransmitter release.

Astrocytes develop from radial glial cells after neuronal differentiation in a particular part of the brain is completed. Radial glial cells will be discussed in Chapter 8, The development of the cerebral cortex. A number of secreted signaling molecules including sonic hedgehog, Wnts, and fibroblast growth factor, as well as a host of as yet unidentified molecules, cause the differentiation of radial glia into astrocytes.

During development, astrocytes play a key role in the formation of active synapses, at least partially due to the formation and release of a family of proteins called thrombospondins. These proteins help form the extracellular matrix (ECM) surrounding synapses. Astrocytes also make and release other ECM components.

Experiments in cell culture demonstrate that neurons cannot form functional synapses without astrocytes. A single astrocyte's processes can surround up to a thousand synapses. During development they can also destroy synapses that are silent. In contrast, they can help enlarge very active synapses so that they can release more neurotransmitter molecules. Astrocytes also play a role in the uptake and degradation of dead cells and ECM components that require destruction. They seem to be able to engulf large amounts of material through a process called phagocytosis. One important protein that they can engulf is the amyloid beta peptide (Aβ). This peptide plays a major role in the etiology of Alzheimer's disease (AD). Aβ's role in AD is discussed in Chapter 18, Trophic factor—receptor interactions that mediate neuronal survival can in some cases last through life, and trophic deficits can produce connectional neurodegenerative diseases.

Astrocytes also play a major role in the formation and maintenance of the blood—brain barrier (BBB) as well as in the functioning of the myelinated axon. These two topics are discussed in a later section of this chapter.

As described in Chapter 4, The growth cone: the key organelle in the steering of the neuronal processes; the synapse: role in neurotransmitter release, astrocytes can form a network of electrically coupled cells. Electrical activation of neurons can stimulate a rapid increase in cytosolic Ca^{++}. This in turn produces a rapid rise in Ca^{++}, via connexons, in a whole network of astrocytes in a wavelike manner. While the precise reason for this propagating Ca^{++} wave is not certain, it definitely suggests that these electrically coupled astrocyte networks are important signaling elements in brain functions.

Oligodendrocytes are the other macroglial cell type in the brain. Their equivalent, Schwann cells, are found in the PNS. Both cells function to wrap their cell membranes around axons to form a tight sheath of myelin, which serves to insulate the axon. This allows the electrical charge to flow at high speed down the axon. A Schwann cell can myelinate only one axon, whereas a single oligodendrocyte can myelinate many. The role of oligodendrocytes and Schwann cells in the evolution, development, and functioning of the myelinated axon will be described in the next section.

As discussed in Chapter 4, The growth cone: the key organelle in the steering of the neuronal processes; the synapse: role in neurotransmitter release, electrical junctions can pass small metabolites from one cell to another. Schwann cells surrounding peripheral nerves can also pass small molecules from one layer of a myelin sheath to another. In the X-linked form of Charcot—Marie—Tooth disease, a mutation in a gene for a single connexin prevents the formation of gap junctions between Schwann cells and leads to demyelination.

Microglial cells constitute the third major glial class and are the major phagocytic cells of the brain (Fig. 6.2). They remove cell debris generated by programmed cell death (apoptosis), synaptic replacement, ECM degradation, etc. Microglia play a critical role in brain development by removing the incorrectly wired neurons and synaptic connections that form during brain maturation.

In contrast to the macroglia, microglia are of mesodermal, not ectodermal, origin. They begin their development in the bone marrow, similarly to that of various cell types in the immune system. Their precursors then migrate into the early developing brain. Microglia have receptors for colony-stimulating factor (CSF), and embryos not containing CSF in their developing brains do not have microglia. Other trophic factor—receptor systems for microglia have not yet been identified in vertebrates. However, members of the PDGF/VEGF family are important in microglia migration into the *Drosophila* brain, and are likely to play a role in this process in the vertebrate brain as well.

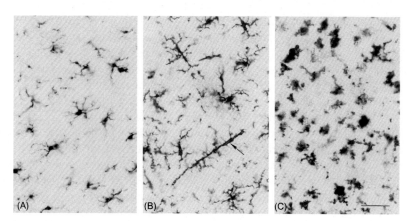

FIGURE 6.2 Activation of microglial cells in a tissue section from human brain.

Resting microglia in normal brain (A). Activated microglia in diseased cerebral cortex (B) have thicker processes and larger cell bodies. In regions of frank pathology (C) microglia transform into phagocytic macrophages, which can also develop from circulating monocytes that enter the brain. Arrow in (B) indicates rod cell. Sections stained with antibody to ferritin. Scale bar = 40 μm.

Squire LR, Berg D, Bloom FB, Du Lac S, Ghosh A, Spitzer NC. Fundamental Neuroscience. 4th ed. Oxford, UK: Elsevier; 2013 [chapters 3, 4, 12], Fig. 3.15, p. 55. With permission.

Microglia enter the brain before the closure of the BBB and remain there until the death of the organism. They can divide and therefore self-renew. The brain is a relatively immunologically protected area and only microglia and astrocytes carry out most immune functions. However, if there is a breech in the barrier, other immune cells can enter. These include T cells and various types of phagocytes such as macrophages and neutrophils.

In synaptic pruning, which occurs during brain development and will be discussed in detail in Chapter 15, Synaptic pruning during development and throughout life depends on trophic factor interactions: the corollary described in Chapter 14, postulating the quantization of trophic factor requirements, provides a plausible explanation for the survival of neurons partially deprived of their full complement of trophic factors, microglia play a key role that involves receptors and their ligands. After synapse formation commences, complement proteins—key components of the immune response—are synthesized and secreted by astrocytes and microglia, and mainly localize on synapses. Synapses that are not active are tagged with complement proteins and recognized by microglial complement receptors. This results in the phagocytosis of the synaptic elements by the microglia.

Microglia in the mature brain are very motile, containing Toll receptors that can recognize a whole array of bacteria and other pathogens. Microglia rapidly produce complements, complement receptors, as well as many cytokines, for example, interleukin 1 and 6, to destroy these pathogens.

Besides continued immune surveillance and mediation of inflammatory responses within the brain, microglia contribute to neuronal function and also to neuronal plasticity throughout life. They engulf apoptotic cells that form in large numbers during brain development using receptors that bind to phosphatidyl serine. This phospholipid is confined to the inner layer of the plasma bilayer except during apoptosis, when it can migrate to the outer layer. Apoptosis will be discussed in detail in Chapter 9, The importance of rapid eye movement sleep and other forms of sleep in selecting the appropriate neuronal circuitry; programmed cell death/apoptosis. The microglia then engulf the apoptotic cell fragments and degrade them in lysosomes. Microglia also synthesize trophic molecules that stimulate neurogenesis throughout life, including insulin growth factor 1, interferon γ, and interleukin 1.

6.2 THE EVOLUTION AND DEVELOPMENT OF THE MYELINATED AXON

As larger organisms developed during evolution, they required longer axons to allow the transmission of electrical current to the presynaptic terminals to escape from predators or to capture prey. Invertebrates for the most part adapted the solution of increasing the diameter of axons that allowed rapid transmission of electrical transmission. One example of an axon with a very large diameter is the squid giant axon that is used to induce rapid movement to escape predators.

Vertebrates adapted a different strategy: to insulate the axon with a multilayered sheath of membranes wound so extremely tight that not even a single proton can escape (Fig. 6.3). Two types of glial cells evolved for this purpose, as mentioned in Section 6.1. Oligodendrocytes are found in the brain and other regions of the CNS and can myelinate many neurons (Fig. 6.4). In contrast, Schwann cells, which myelinate peripheral axons, can myelinate only a single axon (Fig. 6.5).

Most scientists believe that myelin evolved during the same period as the ancestral bony fishes. While there is no fossil record of myelin, there are fossils of bony fishes containing long and narrow openings in the jaw, thought to be locations where myelinated axons were present. These openings are about 10 times as long as those in the jaws of the ancestors of nonbony fish, which are presumed not to have been myelinated.

A key advantage of the myelinated axons present in all vertebrates is that they have much smaller volumes than do the equivalent naked axons in invertebrates, while both conduct electrical charge at the identical speed. Myelination permitted

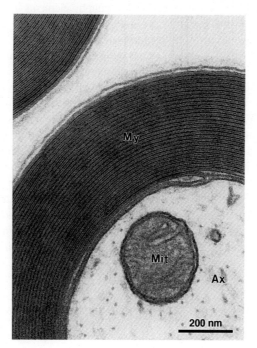

FIGURE 6.3 An electron micrograph of a transverse section through part of a myelinated axon from the sciatic nerve of a rat.

The tightly compacted multilayer myelin sheath (My) surrounds and insulates the axon (Ax). Mit, mitochondria. Scale bar = 200 nm.

Squire LR, Berg D, Bloom FB, Du Lac S, Ghosh A, Spitzer NC. Fundamental Neuroscience. 4th ed. Oxford, UK: Elsevier; 2013 [chapters 3, 4, 12], Fig. 3.7, p. 49. With permission.

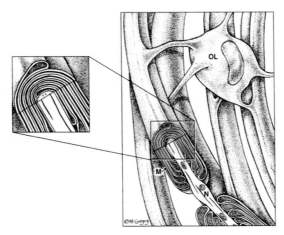

FIGURE 6.4 An oligodendrocyte (OL) in the central nervous system is depicted myelinating several axon segments.

A cutaway view of the myelin sheath is shown (M). Note that the internode of myelin terminates in paranodal loops that flank the node of Ranvier (N). (Inset) An enlargement of compact myelin with alternating dark and light electron-dense lines that represent intracellular (major dense lines) and extracellular (intraperiod line) plasma membrane appositions, respectively.

Squire LR, Berg D, Bloom FB, Du Lac S, Ghosh A, Spitzer NC. Fundamental Neuroscience. 4th ed. Oxford, UK: Elsevier; 2013 [chapters 3, 4, 12], Fig. 3.8, p. 50. With permission.

vertebrates to develop brains that contained many times the number of neurons than those of the equivalent-sized invertebrate brains. This allowed the evolution of structures associated with complex behaviors we call "intelligence." Despite this fact, some invertebrates, for example, octopi, are highly "intelligent" in spite of having orders of magnitude fewer neurons.

Myelin consists of approximately 70% lipid rich in the membrane-stiffening cholesterol, together with 30% protein. The major protein components of the oligodendrocyte myelin membrane are the myelin basic protein (MBP), a cytoplasmic protein made on ribosomes localized in the distal cytoplasm, and the integral membrane proteolipid protein (PLP). PLP functions by linking one myelin layer to another via a homophilic binding to another PLP molecule. These specific interactions can be considered a class of homophilic receptor—receptor interactions. Multiple interactions mediated by PLP—PLP binding cause a very closely connected surface to form between the two opposing plasma membranes, squeezing out all the cytosol except that which contains mainly multilamellar bodies (MLB). Subsequent PLP—PLP interactions between additional plasma membranes lead to a series of very tightly opposed membranes (Fig. 6.6).

Mutations in PLP cause loss of oligodendrocytes and inadequate myelination during development. Autoantibodies against MBP are responsible for multiple

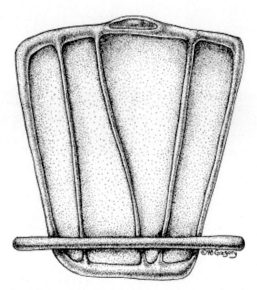

FIGURE 6.5 An "unrolled" Schwann cell in the peripheral nervous system is illustrated in relation to the single axon segment that it myelinates.

The broad stippled region is compact myelin surrounded by cytoplasmic channels that remain open even after compact myelin has formed, allowing an exchange of materials among the myelin sheath, the Schwann cell cytoplasm, and perhaps the axon as well.

Squire LR, Berg D, Bloom FB, Du Lac S, Ghosh A, Spitzer NC. Fundamental Neuroscience. 4th ed. Oxford, UK: Elsevier; 2013 [chapters 3, 4, 12], Fig. 3.9, p. 50. With permission.

sclerosis, which produces the loss of myelin in both the brain and spinal cord and leads to a slowing of the electrical conduction rate. This disease usually begins in young adults.

MBP is also present in the cytosol of Schwann cells. However, instead of the PLPs there are integral plasma membrane proteins called Po. These proteins play the same role as PLPs by binding identical Po molecules on the opposing membrane to make extremely compact membrane layers.

Myelination of axons by both oligodendrocytes and Schwann cells requires precise signaling mechanisms. these require cooperation between developing axons and myelin producing cells as well as the ECM. In the brain and in the PNS, myelination begins after axons have reached their targets (Fig. 6.7).

Oligodendrocytes are the last cell type to differentiate from the radial glia, which are discussed in Chapter 8, The development of the cerebral cortex. Several known signaling ligands including Wnt, neurotrophins, FGF, and insulin like growth factor 1 are secreted by neurons. These trophic factors interact with their receptors, Frizzled/LRP6, Trks, FGFRs, and IGFRs respectively on the oligodendrocyte precursor cell membrane. These binding events signal through the catenin, Akt/Tor, and MAPK/ERK pathways to express genes in the nucleus that

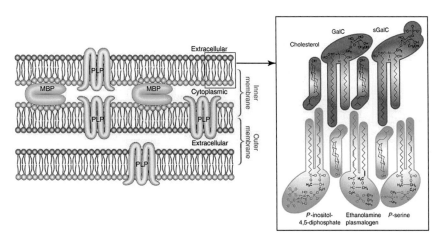

FIGURE 6.6 The structure of the myelin sheath.

Schematic representation of the major myelin components. Compact myelin is formed by the apposition of the external surfaces and internal surfaces of the myelin bilayer that constitute the intraperiodic line and the major dense line, respectively. The myelin bilayer has an asymmetric lipid composition. Some of the major lipid classes are shown on the right.

Aggarwal S1, Yurlova L, Simons M. Central nervous system myelin: structure, synthesis and assembly. Trends Cell Biol. 2011;21:585–593. PMID 21763137. Fig. 2. With permission.

produce proteins necessary for oligodendrocyte differentiation. Initial contact with the axonal membrane and synthesis and distribution of MBP and PLP in the distal cytoplasm and plasma membrane causes differentiation of the oligodendrocyte to occur. This in turn produces concentric membrane elements that form the myelin membrane surrounding a number of discrete axons.

While these events are taking place, numerous processes are occurring in the axons as well, involving myelin–axon interactions. Chief among these interactions is the creation of an axon capable of carrying an electrical current, the action potential, that can travel down a myelinated axon at 100 m/s. To accomplish this task, Na^+ and K^+ channels must be inserted in the axon membrane every 20 μm, the distance that the action potential can travel before dissipating. Therefore, every 20 μm there is a naked axonal membrane called the node of Ranvier, where a large concentration of Na^+ channels is located. This localization is necessary to regenerate the action potential. These channels open to allow rapid Na^+ entry into the cytosol down its concentration gradient. Then the channels rapidly close.

On either side of the node in the paranodal regions, K^+ channels are clustered. After the Na^+ channels close, the K^+ channels open, allowing K^+ ions to flow from inside to outside the axon, down its concentration gradient. This results in the repolarization of the axon after the action potential has passed. Also a high

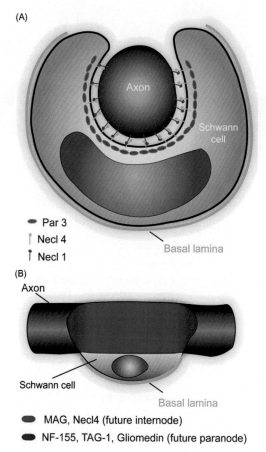

(A)

Par 3
Necl 4
Necl 1

(B)

MAG, Necl4 (future internode)
NF-155, TAG-1, Gliomedin (future paranode)

FIGURE 6.7 Glial–axon recognition and establishment of polarity as early events in myelination in the peripheral nervous system.

(A) The adhesion molecules Necl4 and Necl1 establish glial–axon contact. The asymmetric distribution of the intracellular protein Par-3 and the basal lamina establish a radial axis of polarity. (B) Longitudinal polarity is evidenced by the distribution of the Schwann cell proteins MAG and Necl4 at the axon–glial junction forming the internode, while TAG-1, gliomedin, and NF155 accumulate at the membrane adjacent to the presumptive node of Ranvier.

Simons MI, Trotter J. Curr Opin Neurobiol. *2007;17:533–540. PMID 17923405. Fig. 1. With permission.*

concentration of Na^+, K^+ ATPase molecules is found in the nodes, pumping Na^+ out and K^+ in using the energy derived from adenosine triphosphate (ATP) hydrolysis. The cytoplasmic K^+ then leaks into the exterior of the axonal membrane through resting K^+ channels. This leakage results in the resting potential of the axon, which is about -60 mV.

The development of the nodes of Ranvier depends on interactions between adhesion receptors located on the axonal and the myelin forming plasma membranes. Knowledge of the protein interactions necessary to form a node of Ranvier in the CNS is incomplete. However, there is evidence for the involvement of at least several classes of receptors in the formation of the nodes of Ranvier in peripheral axons. Gliomedin and NrCAM on the Schwann cell surface bind to the axonal membrane protein neurofascin 186 by a mechanism analogous to a receptor−receptor interaction (Fig. 6.8). A number of these binding events demarcate the boundaries of each node.

The Na^+ and K^+ channel molecules and other molecules that will eventually end up in the nodes are initially dispersed throughout the axon. After the binding

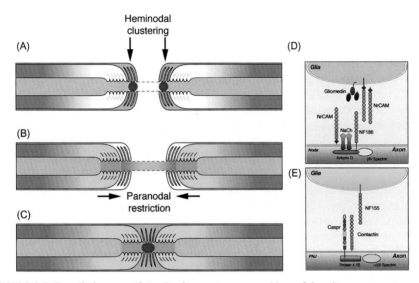

FIGURE 6.8 Two distinct axoglial adhesion systems assemble peripheral nervous system nodes of Ranvier.

(A) Na^+ channels (*red circle*) are trapped at heminodes that are contacted by Schwann cell microvilli (MV; *orange*). Axon−glia interaction at this site is mediated by binding of gliomedin and glial NrCAM to axonal NF186 (D). A transmembrane and secreted forms of glial NrCAM trap gliomedin on Schwann cell microvilli and enhances its binding to NF186. In the absence of either gliomedin or glial NrCAM, Na^+ channels fail to cluster at heminodes. Binding of gliomedin to axonal NrCAM, which lacks its cytoplasmic domain, does not result in Na^+ channel clustering. (B) The distribution of Na^+ channels is restricted between two forming myelin segments by the PNJ (*blue*). Three CAMs, NF155 present at the paranodal loops (*green*), and axonal complex of Caspr and contactin mediate axon−glia interaction and the formation of the PNJ (E). (C) These two cooperating mechanisms provide reciprocal backup systems and ensure that Na^+ channels are found at high density at the nodes. *PNJ*, paranodal axoglial junction.

Feinberg K, et al. Neuron. 2010;65:490−502. Fig. 8. PMID: 20188654. With permission.

of neurofascin 186 to NrCAM and gliomedin, the cytosolic protein, ankyrin, in the axonal cytoplasm, together with other peripheral membrane molecules, tether the Na^+ and K^+ molecules and the Na^+, K^+ ATPase molecules to form a node of Ranvier (Fig. 6.8). After this process occurs the Na^+ channel is concentrated 100 times in the node of Ranvier compared with its concentration in nonmyelinated axons (Fig. 6.8).

At the nodes of Ranvier, a tight apposition of several cell types occurs, forming a structure analogous to a synapse. The oligodendrocyte supplies necessary nutrients for ATP production including lactate. There is a capillary that releases oxygen and glucose into the extracellular space. These substances are taken up by the axon, which requires a high level of ATP to maintain the ionic gradients of the axon. The oxygen can diffuse into the axon. Glucose, however, is transported down its concentration gradient by a member of glucose transporter family, Glut3, which is only present on neurons. In this tightly enclosed extracellular space there is also an astrocyte, which induces the CEC to form extremely tight junctions. These in turn form the BBB. The astrocytes presumably accomplish this task by secreting specific ligands recognized only by the receptors on the CEC surface. Finally, a cell type called a pericyte is present. This cell wraps around the endothelial membrane and secretes proteins that form the ECM. The BBB will be discussed at length in the next section.

A very interesting question that is still unanswered is: How did the synapse-like structure at each node of Ranvier evolve? The production of this structures requires great coordination between many developing cell types. Therefore, one finds it hard to imagine how the long myelinated axon, with its great energy demands that cannot be met by the nerve cell body, could have evolved in stages. Perhaps the first myelinated axon was less than 20 μm long, thus not requiring a node of Ranvier with its necessary additional protein requirements. At later times the axons grew longer and additional proteins were added to form nodes of Ranvier.

6.3 BLOOD—BRAIN BARRIER

Once a significant nervous system evolved, there was a requirement for a cellular barrier separating the CNS from the periphery. This requirement is essential for a number of reasons. One is to protect the brain from toxins that are lethal to brain cells, particularly neurons, which for the most part cannot be replaced once they die. A second major reason is to maintain a very constant ionic composition in the extracellular space surrounding synapses, which is required for the synapses' many functions. Thirdly the brain must be an immune protected organ, free from invasion by immune cells from the periphery. Finally, as described in the last section, capillaries are required to deliver oxygen, glucose, and other substances to the nodes of Ranvier of myelinated axons as well as to other parts of the brain.

In invertebrates the barrier between periphery and brain is maintained by a glial barrier. However, with the evolution of vertebrates, the barrier now includes several types of cells. One is a special class of endothelial cell called the CEC, which forms the primary barrier. These cells require the end feet of astrocytes that contact the brain-facing side of the endothelial cell surfaces to produce the necessary tight junctions. One other cell type is important for BBB development and maintenance. As described in the previous section, pericytes almost completely surround the CECs on the brain side and help induce their barrier properties. They also secrete a group of ECM components that help form a basement lamina between the pericytes and the astrocyte end feet (Fig. 6.9). Pericytes also contain a network of contractile proteins in their cytoplasm that behaves similarly to that in the smooth muscle cells that surround endothelial cells in other areas of the body by rendering a contractile tone to the BBB.

The BBB forms extremely tight junctions. These tight junctions almost totally prevent the passage of hydrophilic molecules and even small charged ions from one side of the barrier to the other. However, hydrophobic molecules, including drugs for the treatment of a variety of brain diseases, can enter the brain by moving across the two bilayers that compose the blood and brain surfaces of the CECs.

The molecules that compose the tight junctions of the capillary endothelium are two classes of integral membrane proteins called claudins and occludins respectively. These molecules specifically interact with one another and can be considered adhesion receptors (Fig. 6.9).

Another distinctive characteristic of CECs is that they are almost devoid of intracellular vesicles, which transfer molecules from one side of the cell to another by a process called transcytosis. However, when transcytosis does occur, it involves the receptor-mediated endocytosis from the blood of a very select group of molecules. These include transferrin bound to its receptor, which delivers the essential iron molecule to the brain. As described in Chapter 1, Classes of receptors, their signaling pathways, and their synthesis and transport, iron is critical for many reasons including the synthesis of ATP in mitochondria.

Insulin and amylin, peptides secreted by the pancreas; are bound to their respective receptors are also delivered to the brain by transcytosis. However, it is unclear why the transport of insulin and amylin into the brain is necessary, because both peptides are synthesized by brain cells.

The BBB endothelial cells also contain a group of receptors for small molecules. Perhaps the most important of these is glucose, the major energy nutrient of neural cells. A high concentration of the Glut 1 glucose transporter is located on the blood side of the BBB, serving to translocate glucose from the blood to the brain. This transport through a channel in Glut 1 does not require energy, because the concentration of glucose in the blood is always higher than that in the brain. The CEC is also totally permeable to the other major energy source of the brain, oxygen. The CECs are also permeable to carbon dioxide, the gaseous waste product of energy production, which leaves the brain. Also small fatty acids,

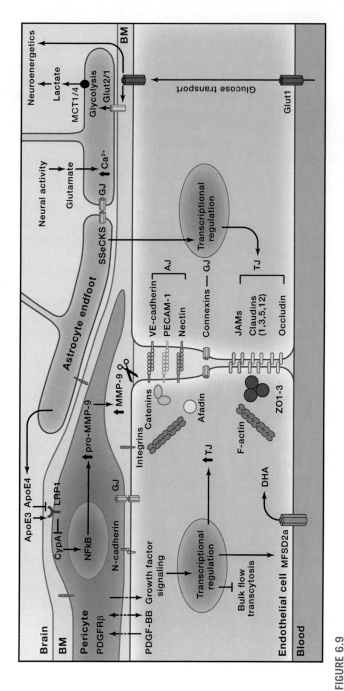

FIGURE 6.9

The vascular triad of the blood–brain barrier. Endothelial cells are connected with each other through the tight junction (TJ), adherens junction (AJ), and gap junction (GJ) proteins. In the TJs, occludin, claudins, and junctional adhesion molecules (JAMs) form an impermeable barrier to fluids and are connected to F-actin filaments by the zonula occludens ZO-1, ZO-2, and ZO-3 multidomain scaffolding proteins of the membrane-associated guanylate kinase family. GJs formed by connexin hemichannels are specialized for direct intercellular communications. AJs are formed by homotypic binding of VE-cadherin, platelet endothelial cell adhesion molecule-1 (PECAM-1), and Nectin. Catenins link VE-cadherin to

(Continued)

F-actin, while nectin is secured to F-actin by afadin. Pericytes communicate with the endothelial cells via growth factor–mediated signaling (unidirectional or bidirectional), adhesion via N-cadherin homotypic binding, and GJs. Pericytes can modulate blood–brain barrier permeability by regulating gene expression in the endothelial cells resulting in upregulation of TJ proteins, inhibition of bulk flow transcytosis, and upregulation of brain endothelial specific docosahexaenoic acid (DHA) transporter, a major facilitator domain-containing protein 2A (MFSD2a). Both endothelial cells and pericytes are embedded in the basement membrane (BM) and anchored to BM via integrins. *PDGF-BB*, platelet-derived growth factor BB; *PDGFRβ*, platelet-derived growth factor receptor-β. Astrocytes regulate expression of matrix metallo-proteinase-9 (MMP-9) in pericytes by secreting apolipoprotein E (ApoE). ApoE3, but not ApoE4, binds to the low-density lipoprotein receptor-related protein 1 (LRP1) in pericytes, which suppresses the proinflammatory cyclophilin A (CypA)-nuclear factor-κB (NFκB)-MMP-9 pathway and degradation of TJ and BM proteins causing BBB breakdown. Astrocytes signal endothelial cells by Src-suppressed C-kinase substrate (SSeCKS) to increase TJ protein expression. Neural activity–dependent glutamate release increases [Ca2 +] in the astrocytic endfeet, which regulates vascular tone. The GJs connect the adjacent astrocytic endfeet. Glucose gets into the brain via the endothelial Glut1 transporter and is taken up by neurons via Glut3. Glucose is taken up by astrocytes mainly by Glut2 and is metabolized to lactate, which is exported to neurons by the monocarboxylic MCT1 and MCT4 transporters.

Zhao Z, Nelson AR, Betsholtz C, Zlokovic BV. Establishment and dysfunction of the blood-brain barrier. Cell. 2015;163:1064–1078. Fig. 3. PMID: 26590417. With permission.

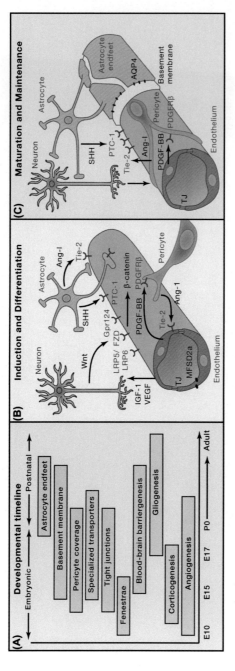

FIGURE 6.10 Blood—brain barrier development in the murine central nervous system.

(A) Developmental timeline. Restriction of paracellular and transcellular transport of solutes is accomplished by elimination of endothelial fenestrae and pinocytosis, formation of a continuous endothelial monolayer connected with the tight junctions, creation of highly selective endothelial transport systems, and establishment of specialized perivascular structures, including the basement membrane and the coverage of the endothelial capillary wall by pericytes and astrocytic endfeet. *E*, embryonic days; *P*, postnatal days. (B) Induction and differentiation. Wnt ligands (Wnt7a/7b) secreted by neural cells bind to endothelial Frizzled receptors (FZD) and the coreceptors low-density lipoprotein receptor—related protein (LRP) 5 and 6, which activate β-catenin signaling, leading to the induction of BBB specific genes. G protein coupled receptor 124 (Gpr124) coactivates Wnt/β-catenin signaling. Endothelial cells secrete platelet-derived growth factor BB (PDGF-BB), which interacts with platelet-derived growth factor receptor-β (PDGFR-β) in pericytes, inducing pericyte recruitment. Pericytes and astrocytes secrete angiopoietin-1 (Ang-1), which acts on endothelial Tie-2 receptor leading to microvascular maturation and highly stable and impermeable blood—brain barrier (BBB). Pericytes are required for the expression of endothelial major facilitator superfamily domain-containing protein 2a (MFSD2a), which is critical for the BBB formation and maintenance. Astrocytes secrete sonic hedgehog

(*Continued*)

(SHH), which acts on endothelial patched homolog 1 (PTC-1) receptor eliciting signaling, which contributes to the BBB formation. Endothelial cells secrete vascular growth factor (VEGF) and insulin growth factor (IGF-1) contributing to proper neurovascular patterning. Additional signal transduction pathways may also participate in BBB formation. *TJ*, tight junction. (C) Maturation and maintenance. Postnatally, brain capillaries are covered by mature pericytes sharing the basement membrane with endothelium. Astrocytic endfeet form the outer layer of the mature capillaries. Pericytes and astrocytes continue secreting matrix proteins (*yellow*) of the basement membrane. Signaling pathways mediating BBB induction and differentiation likely continue to play a role in BBB maturation and maintenance and their dysregulation may lead to BBB breakdown causing different central nervous system pathologies. *AQP4*, aquaporin-4 water channel.

Zhao Z, Nelson AR, Betsholtz C, Zlokovic BV. Establishment and dysfunction of the blood-brain barrier. Cell. 2015;163:1064–1078. Fig. 2. PMID: 26590417. With permission.

which under some conditions can serve as an energy source for neural cells, can cross the BBB because of their hydrophobicity.

The CECs also contain a group of transporters on the brain-facing side of the CECs. They bind to and transport potentially toxic molecules out of the brain. They also contain receptors that bind to and transcytose the Aβ out of the brain. Aβ is thought by many investigators to be the cause of AD. Aβ and its role in AD will be discussed in Chapter 18, Trophic factor−receptor interactions that mediate neuronal survival can in some cases last through life, and trophic deficits can produce connectional neurodegenerative diseases.

The development of the BBB is a highly controlled process that intimately depends on signals transmitted between the precursors to the abovementioned cells via ligand−receptor interactions. Initially angioblasts, the precursors to endothelial cells, penetrate the neuroectoderm using the guidance cue of vascular endothelial growth factor binding to its receptor. Wnt signaling by neural cells to the leaky endothelial cells containing Frizzled/Lrp6 coreceptors is a key event in inducing genes necessary for BBB formation, including Glut 1 (Fig. 6.10).

An unanswered question, similar to that posed regarding the evolution of the myelinated axon, is: How did the BBB evolve? Again, the development of this complex of different cell types required great coordination of signals as discussed previously. Is it possible that this structure evolved in stages, or did the formation of the BBB evolve all at once?

RECOMMENDED BOOK CHAPTERS

Kandel ER, Schwartz JH, Jessel TM, Siegelbaum SA, Hudspeth AJ. *Principles of Neural Science*. 5th ed. New York: McGraw Hill; 2013 [chapters 4, 7, Appendix D. Also recent review articles].

Squire LR, Berg D, Bloom FB, Du Lac S, Ghosh A, Spitzer NC. *Fundamental Neuroscience*. 4th ed. Oxford, UK: Elsevier; 2013. chapters 3, 4, 12.

Review Articles

Aggarwal S1, Yurlova L, Simons M. Central nervous system myelin: structure, synthesis and assembly. *Trends Cell Biol*. 2011;21:585−593.

Zhao Z1, Nelson AR1, Betsholtz C2, Zlokovic BV. Establishment and dysfunction of the blood-brain barrier. *Cell*. 2015;163:1064−1078.

A testable theory on the wiring of the brain

7

CHAPTER OUTLINE

Hypothesis 1: Early in development each presumptive neuron expresses a unique set of receptors and trophic factors (Fig. 7.1A). Once each neuron produces its particular receptors, it becomes dependent on a continuous supply of the trophic factors that bind to these receptors. These trophic factors provide the necessary ingredients to sustain its life; otherwise the neuron will die. The receptor choices each neuronal precursor makes result in it having a unique anatomy and functions. A variety of neurotropic and repulsive proteins are used to guide each neuronal process containing a specific receptor class to its appropriate target (Fig. 7.1B). This topic will be the focus of much of the remainder of this book.

There is already substantial evidence supporting this hypothesis. Over many years, experiments have demonstrated the existence of neurotrophic and repulsive molecules in both cell culture and in vivo models.

Corollary 1: There is a quantization of trophic factor requirements. For example, if a neuronal precursor expresses three quanta of a particular receptor, it will require three quanta of the trophic factor to which it binds. Each quantum of a given receptor is likely found on a different process (Fig. 7.1A and B).

To my knowledge, no experiments supporting this corollary have been reported. However, it should be testable using newly developed techniques. These should allow one to measure the number of specific receptor molecules on a single neuron in cell culture. It should also be possible to determine the localization of each quantum of a given receptor on either cell body, dendrite, or axon.

One major advantage, should this corollary be true, is that the number of specific receptor−trophic factor combinations required to wire the brain correctly would be significantly fewer. This corollary is also useful in explaining the molecular mechanism of synaptic pruning and synaptic plasticity, key events in organizing the brain during development and throughout adult life. These topics are examined in detail in Chapter 15, Synaptic pruning during development and

Receptors in the Evolution and Development of the Brain. DOI: https://doi.org/10.1016/B978-0-12-811012-6.00007-8

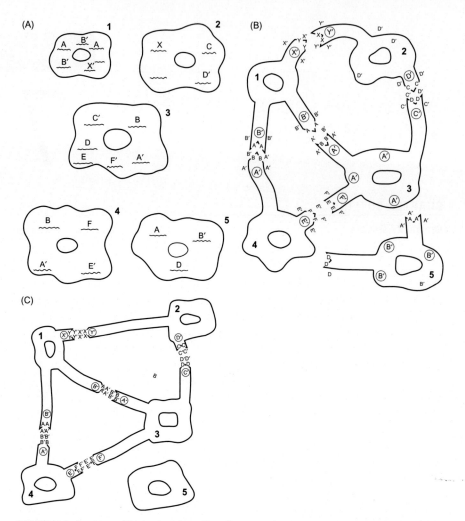

FIGURE 7.1 A cartoon illustrating the salient features of the hypothesis emphasizing the key role of receptor–trophic factor interactions in brain development

(A) Early in development mRNAs for receptors and trophic factors are produced. This panel illustrates this point. Five cells producing mRNA for receptors A B C D E F G H and trophic factors A′ B′ C′ D′ E′ F′ G′ H′ are shown here. The quantization hypothesis is illustrated by the requirement of cell 1 for two quanta of A′ because of its production of two quanta of mRNA for A. (B) The cells, which are now committed to requiring the trophic factors that bind to the receptors they express, send out processes with receptors located at the growth cones. These seek connections with a cell or cells secreting complementary factors. We visualize secretion of trophic factors occurring from growth cones of axons and/or dendrites as well as from cell bodies. (C) Cells 1–4 have made appropriate connections. Functional activation of synaptic connections has resulted in appropriate trophic requirements for these cells being met. Cell 5 has made one appropriate connection. However, it has not found a cell to supply its need for trophic factor D′ and hence, does not survive. Note that most of the connections pictured here are reciprocal. We do not, however, mean to imply that this represents the in vivo situation.

Fine RE, Rubin JB. Specific trophic factor-receptor interactions: key selective elements in brain development and "regeneration". J Am Geriatr Soc. 1988;36:457–466, Fig. 1. PMID: 2834427. With permission.

throughout life depends on trophic factor interactions: the corollary described in Chapter 14, postulating the quantization of trophic factor requirements, provides a plausible explanation for the survival of neurons partially deprived of their full complement of trophic factors.

Hypothesis 2: After a portion of the brain is wired, the neurons are caused to fire constantly during rapid eye movement (REM) sleep, as well as to a lesser degree during slow wave sleep. The neurons that have formed the correct synaptic connections will receive their required trophic factors while the incorrectly wired ones will not (Fig. 7.1C). This topic will be extensively discussed in Chapter 9, The importance of rapid eye movement sleep and other forms of sleep in selecting the appropriate neuronal circuitry; programmed cell death/apoptosis.

To my knowledge there has been no concrete connection made between REM sleep and selection of the properly wired neurons, other than the article published by Fine and Rubin (see General References at the end of this chapter). However, it would appear to be self-evident based on much evidence indicating that there is significantly increased neurotransmitter release during REM sleep in many regions of the developing brain.

Hypothesis 3: Improperly wired neurons will die by apoptosis (Fig. 7.1C). The resulting residual vesicles will be engulfed by microglia and degraded in lysosomes. This topic will also be examined in detail in Chapter 9, The importance of rapid eye movement sleep and other forms of sleep in selecting the appropriate neuronal circuitry; programmed cell death/apoptosis.

There is substantial evidence supporting this hypothesis.

Hypothesis 4: Connectional developmental and age-related neurodegenerative diseases are produced at least in part by deficits in receptor—trophic factor interactions. "Connectional" describes neurons that communicate with one another via synapses. These topics will be considered in detail in Chapter 17, Role of trophic factors and receptors in developmental brain disorders, and Chapter 18, Trophic factor—receptor interactions that mediate neuronal survival can in some cases last through life, and trophic deficits can produce connectional neurodegenerative diseases.

There is already a great deal of evidence supporting this hypothesis. However, it hasn't been presented in the context of the above hypotheses except in the Fine and Rubin article (see General References).

Hypothesis 5: Neuronal stem cell transplants offer a possible cure for the above-mentioned diseases by employing receptor—trophic/tropic factor interactions. This will be the topic of discussion in Chapter 19, Potential use of neuronal stem cells to replace dying neurons depends on trophic factors and receptors.

There is already evidence that neuronal stem cell transplants are successful in animal models. However, to my knowledge there has not been a satisfactory theory that explains their success, in light of the hypotheses put forward in this chapter.

Once the organization of the brain occurs, the neural precursor cells begin to differentiate. Then their cell bodies, primitive dendrites, and axons migrate to their final destinations. The migration routes of the two major cortical neuron types, the glutamate (Glu) and the γ aminobutyric acid (GABA) containing neurons, are described in the next chapter. Neuronal precursors in other brain regions must also migrate, put out processes, and ultimately connect with the correct partner. This requires each neuronal precursor to find its way through a veritable jungle of other migrating cells and processes.

At the time that early neuronal precursors are determined to become neurons and the precursors have multiplied, each one expresses a unique array of receptors and ligands. This choice probably depends on many factors, including where in the developing brain the cell is located, who its neighbors are, and the precise time in which this choice occurs. Whatever factors determine each cell's decision, it is very likely that each neuronal precursor in the mammalian brain makes both qualitative and quantitative decisions. This choice very likely results in every neuron in the brain being unique in its precise anatomy and function. This possibility has not been explored in neurons. However, in cell culture, fibroblasts cloned from a single cell express wide qualitative and quantitative differences in the expression of different mRNAs and in the proteins that they produce. If this is true of fibroblasts, which are a very "simple" cell, it must be the case with neurons, the most complex cells in the body.

The ability of the receptors and ligands to influence specific developmental patterns in neuronal identity, migration, synaptogenesis, and ultimately survival is the key element in this theory. The ligands can be secreted molecules including small neurotransmitters such as Glu, GABA, dopamine, serotonin, or acetylcholine; small neuropeptides including vasopressin, substance P, neuropeptide Y; and larger trophic molecules including nerve growth factor and other members of the neurotrophin family. There are also attractive or inhibitory molecules including ephrins, netrins, and semaphorins. These molecules are either secreted and/or integral plasma membrane components. Using a whole array of neurotropic, neurotrophic, and inhibitory interactions, the growth cone of a unique neuronal process locates another growth cone having the complementary receptors on its plasma membrane. These growth cones then become a synapse (Fig. 7.1C).

Fig. 7.1C also indicates that there are reciprocal trophic connections between neurons, with trophic substances being secreted from postsynaptic endings as well as from presynaptic endings. There is evidence that this occurs in the case of neurotrophins. The postsynaptic endings secrete the neurotrophin that is bound to the presynaptic ending's Trk receptors, endocytosed, and retrogradely transported down the axon to the cell body. Signals from the endocytosed vesicles activate nuclear transcription factors. These in turn activate genes producing mRNAs and finally proteins that allow survival. The neurotrophin signaling pathway is presented in detail in Chapter 10, The key roles of brain-derived growth factor and endocannabinoids at various stages of brain development including neuronal commitment, migration, and synaptogenesis.

Another group of factors released from the postsynaptic ending are the cannabinoids. As discussed in Chapter 10, The key roles of brain-derived growth factor and endocannabinoids at various stages of brain development including neuronal commitment, migration, and synaptogenesis, cannabinoids are lipids that are synthesized and released into the synaptic cleft and bound to presynaptic receptors. The cannabinoid receptors are members of the G-protein coupled receptor (GPCR) superfamily and are the most abundant group of GPCRs in the brain. Cannabinoids play mostly inhibitory roles in neurotransmitter release and interact with neurotrophin receptors during development and throughout life.

Following the discussion of cerebral cortex development in the next chapter, other chapters will be devoted to topics related to the role of receptors in various aspects of brain development. The mechanisms by which receptor—trophic factor interactions play key roles in drug addiction, viral infections, and developmental and age-related connectional diseases are described. The role of trophic factors and receptors in the development of the enteric nervous system will be examined, as well as the critical role of the microbiota in the development of both the enteric nervous system and the brain. The important topic of synaptic pruning during development, and the role of trophic factor—receptor interactions in this process, will also be described. Finally, the possibility of curing connectional diseases through stem cell transplants, which rely on trophic interactions to reestablish the proper circuitry, will be discussed.

GENERAL REFERENCES

Fine RE, Rubin JB. Specific trophic factor-receptor interactions: key selective elements in brain development and "regeneration". *J Am Geriatr Soc*. 1988;36:457−466.

Lawrence J, Singer R. Intracellular localization of messenger RNAs for cytoskeletal proteins. *Cell*. 1986;45:407−415.

Development of the cerebral cortex

CHAPTER OUTLINE

As discussed in the previous chapter, how the mammalian brain accomplishes the selection of the proper wiring will be a major theme for the remainder of this book. However, before proceeding with this topic, the formation of the most complex portion of the brain, the cerebral neocortex, will be examined.

Cortical development illustrates the complexity of creating the mammalian brain. The cerebral neocortex is the newest addition to our brain from an evolutionary and developmental standpoint, and is the biological substratum of human mental prowess. While its origin appears to be in reptiles, the cortex first appeared as a uniform six-layered structure of neurons and associated glial elements in small mammals, which began to evolve between 260,000 and 215,000 years ago. The neocortex evolved in early mammals and is the largest component in the mammalian brain. Its size varies by more than 1000-fold between species. Primates, especially humans, have the largest neocortex compared with their body weight. The human cerebral cortex is three times as large as its nearest relative, the chimpanzee. This great expansion is almost solely due to the much increased size of the prefrontal cortex, the center of cognition and decision making.

The increased size of the primate cerebral cortex is due the large number of folds in the lateral dimension when the radial dimension remains almost constant. These folds or sulci allow for a tremendous increase in the number of cells in every region of the neocortex without the necessity of increasing skull size proportionately. Rodents, in contrast, have an almost flat cortical surface (Fig. 8.1).

The neurons of the neocortex can be divided into two major classes: excitatory and inhibitory. The development of each significantly depends on trophic and tropic interactions. The projection neurons comprising the six laminar layers contain the excitatory transmitter, Glu. Layer I is the outermost, nearest to the pial surface, and VI is the innermost, nearest to the ventricle. Excitatory projection neurons make up 70%−80% of the cortical neurons. Their myelinated axons project to other neurons on the same side of the cortex, to neurons in the other half of the cortex, as well as to neurons in other brain regions (Fig. 8.2). Pyramidal cells (another name for the cortical projection neurons because of the shape of their cell bodies) have a large number of dendrites but only one axon. This is true for

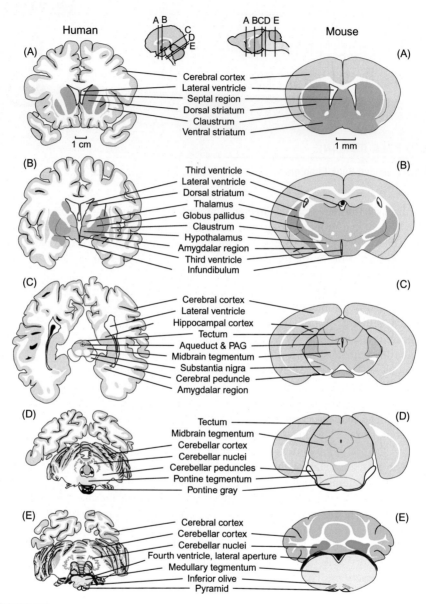

Human   Mouse

(A)
- Cerebral cortex
- Lateral ventricle
- Septal region
- Dorsal striatum
- Claustrum
- Ventral striatum

1 cm   1 mm

(B)
- Third ventricle
- Lateral ventricle
- Dorsal striatum
- Thalamus
- Globus pallidus
- Claustrum
- Hypothalamus
- Amygdalar region
- Third ventricle
- Infundibulum

(C)
- Cerebral cortex
- Lateral ventricle
- Hippocampal cortex
- Tectum
- Aqueduct & PAG
- Midbrain tegmentum
- Substantia nigra
- Cerebral peduncle
- Amygdalar region

(D)
- Tectum
- Midbrain tegmentum
- Cerebellar cortex
- Cerebellar nuclei
- Cerebellar peduncles
- Pontine tegmentum
- Pontine gray

(E)
- Cerebral cortex
- Cerebellar cortex
- Cerebellar nuclei
- Fourth ventricle, lateral aperture
- Medullary tegmentum
- Inferior olive
- Pyramid

FIGURE 8.1

Mini atlases to compare major adult brain regions in humans and mice. The brains are cut approximately transversely to the CNS longitudinal axis and illustrate five major levels, arranged from rostral to caudal: (A) endbrain, (B) interbrain, (C) midbrain, (D) pons, and (E) medulla. The choroid plexus of the lateral, third, and fourth ventricles shown in red.

Adapted from Squires LR, Berg D, Bloom FB, Du Lac S, Ghosh A, Spitzer NC. Fundamental Neuroscience.
4th ed. Oxford: Elsevier; 2012 [Chapters 13–15, 50]. Fig. 2.15, p. 28. With permission.

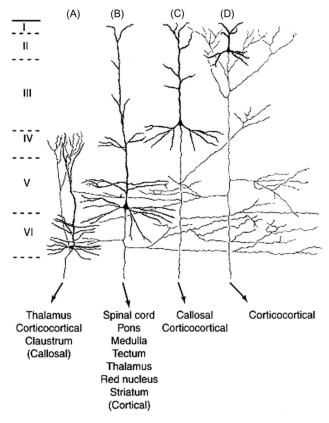

                    (A)   (B)   (C)   (D)

Thalamus   Spinal cord   Callosal   Corticocortical
Corticocortical   Pons   Corticocortical
Claustrum   Medulla
(Callosal)   Tectum
                   Thalamus
                  Red nucleus
                   Striatum
                   (Cortical)

FIGURE 8.2

Morphology and distribution of neocortical pyramidal neurons. Note the variability in cell size and dendritic arborization, as well as the presence of axon collaterals, depending on the laminar localization (I–VI) of the neuron. Also, different types of pyramidal neurons with a precise laminar distribution project to different regions of the brain (less important projection zones are indicated by parentheses).

Squires LR, Berg D, Bloom FB, Du Lac S, Ghosh A, Spitzer NC. Fundamental Neuroscience. 4th ed. Oxford: Elsevier; 2012 [Chapters 13–15, 50]. Fig. 3.6, p. 48. With permission.

all other neurons in the central nervous system. The Purkinje cells of the cerebellum are an extreme example, having an enormous number of dendrites (Fig. 8.3).

The interneurons, which contain the inhibitory neurotransmitter GABA, comprise the other 20%–30% of the cortical neurons, and modulate the firing of the excitatory projection neurons. They are interspersed throughout the six laminar areas of the cerebral cortex. This class of neurons has short unmyelinated axons and many dendrites, and form synapses locally (Fig. 8.4).

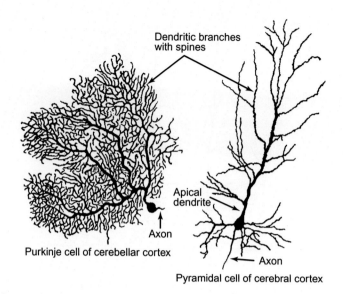

Dendritic branches
with spines

Apical
dendrite

Axon

Purkinje cell of cerebellar cortex

Axon

Pyramidal cell of cerebral cortex

FIGURE 8.3

Typical morphology of projection neurons. (Left) A Purkinje cell of the cerebellar cortex and (right) a pyramidal neuron of the neocortex. These neurons are highly polarized. Each has an extensively branched, spiny apical dendrite, shorter basal dendrites, and a single axon emerging from the basal pole of the cell. The scale bar represents approximately 200 μm.

Squires LR, Berg D, Bloom FB, Du Lac S, Ghosh A, Spitzer NC. Fundamental Neuroscience. 4th ed. Oxford: Elsevier; 2012 [Chapters 13–15, 50]. Fig. 3.1, p. 42. With permission.

The cytoarchitecture of the mammalian neocortex is a product of these two major classes of cortical neurons. The neurons are interspersed and supported by several types of glial cells and blood vessels that together form a thin mantle of gray matter. The gray matter overlays the white matter, which is mainly composed of myelinated axons from the Glu-containing excitatory neurons.

In the radial dimension the cerebral cortical gray matter is composed of columns of nerve cells, which function together as modular units. In each column the cells are interconnected in an identical fashion. Each column is also connected to other columns. Large groups of columns in different parts of the neocortex perform different functions, for example, the visual cortex is involved in sight while the motor cortex is involved in movement (Fig. 8.5).

The Glu- and GABA-containing classes have different anatomical origins. A remarkable feature of the development of the neocortex is that none of its constituent neurons are formed in the cortex itself.

At an early time in embryogenesis, a small group of cells in the primitive neural tube are determined to become neural cells by the interaction of a number of trophic factors with their specific receptors. Among these molecules are the Wnts, members of

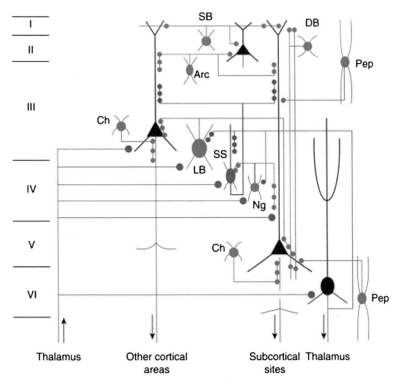

FIGURE 8.4 A schematic circuit based upon the known cortical cells upon which thalamic afferent fibers terminate in cats and monkeys.

The GABAergic interneurons (*blue*) are identified by the names that they have received in these species. *Arc*, neuron with arciform axon; *Ch*, chandelier cell; *DB*, double bouquet cell; *LB*, large basket cell; *Ng*, neurogliaform cell; *Pep*, peptidergic neuron; *SB*, small basket cell. Excitatory neurons (*black*) include pyramidal cells of layers II—VI and the spiny stellate cells (SS) of layer IV.

Squires LR, Berg D, Bloom FB, Du Lac S, Ghosh A, Spitzer NC. Fundamental Neuroscience. 4th ed. Oxford: Elsevier; 2012 [Chapters 13–15, 50]. Fig. 3.5, p. 46. With permission.

the TGFβ family, fibroblast growth factors (FGFs), and sonic hedgehog (SHH) (see Chapter 1: Classes of receptors, their signaling pathways, and their synthesis and transport, and Chapter 7: A testable theory to explain how the wiring of the brain occurs).

After trophic factor—receptor binding occurs, signaling pathways to the cell nucleus are stimulated, ultimately leading to the activation of specific transcription factors. These transcription factors acting via receptor—ligand interactions between cells, specifying the early organization of the brain into various segments including the forebrain, midbrain, hind brain, and spinal cord. The cerebral cortex develops from the forebrain, the thalamus and striatum form the midbrain, and the cerebellum and brainstem, including the pons and medulla form the hind brain.

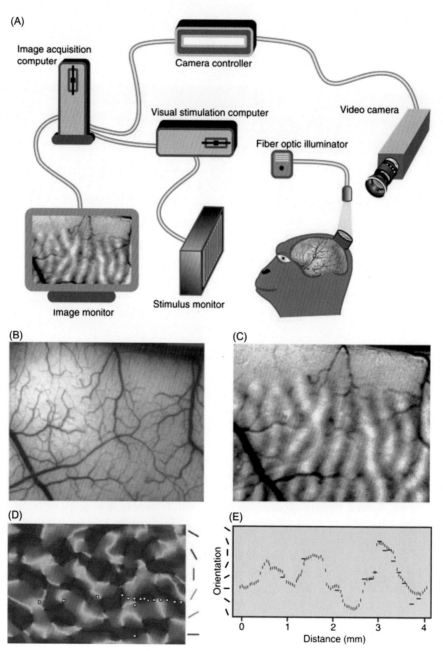

FIGURE 8.5 Optical imaging of functional architecture in the primate visual cortex.

(A) Schematic diagram of the experimental setup for optical imaging. Digitized images of a region of visual cortex (as in B) are taken with a CCD camera while the anesthetized,

(Continued)

The cerebral cortex derives from a group of morphologically identical cells classified as the neuroepithelium. Different parts of this structure become different regions of the cortex. For example, one region becomes the hippocampus, another the neocortex. There are further subdivisions of each region. The rostral region of the neocortex becomes the motor and frontal cortices, while caudal regions are involved in sensory perception.

Trophic factors and receptors play key roles in the early boundary formation and division of the cortex. A group of cells called the rostral patterning center supplies positional information along the rostral—caudal axis. Forebrain patterning is determined by FGF signaling interacting with a dorsal center. This center then secretes members of the BMP and Wnt families. In contrast, the ventral center secretes SHH. Interactions between these secreted molecules and their receptors confer three-dimensional positional identity to each neuronal precursor in the early forebrain (Fig. 8.6).

All of the excitatory neurons in the cerebral cortex derive from a small number of neural epithelial cells (NECs), which are polarized in an apical—basal direction with respect to the ventricle surface. Adherens junctions, composed of members of the cadherin receptor family and other proteins, tightly connect the cells to each other. There are also gap junctions and tight junctions between adjacent cells (Fig. 8.7). NECs constitute the early neural tube. These cells initially divide symmetrically producing expansion in the lateral dimension.

The origin of the stem cell precursors is the transient proliferative zone also called the ventricular zone (VZ), situated on top of the surface of the cerebral lateral ventricles in the middle of the brain. Neurogenesis begins with some NECs switching to asymmetric, differentiating cell divisions. One cell in each asymmetric division becomes a neuronal precursor cell and the other remains a proliferating cell called a neuronal stem cell. These divisions result in the linear increase in

---

◀ paralyzed animal is viewing a visual stimulus. These images are stored on a second computer for further analysis. (B) Individual image (9 × 6 mm) of a region of V1 and a portion of V2 taken with a special filter so that blood vessels stand out. (C) Ocular dominance map. Images of the brain during right-eye stimulation were subtracted digitally from images taken during left-eye stimulation. (D) Orientation map. Images of the brain were taken during stimulation at 12 different angles. The orientation of stimuli that produced the strongest signal at each pixel is color coded, as indicated to the right. The key at the right gives the correspondence between color and the optimal orientation. (E) Comparison of the preferred orientation of single neurons with the optical image. At each of the locations indicated by squares in D, single neurons were recorded with microelectrodes. The preferred orientations of the neurons (*dashes*) were compared with the preferred orientations measured in the optical image, sampled along the line connecting the recording sites (*dots*).

Squires LR, Berg D, Bloom FB, Du Lac S, Ghosh A, Spitzer NC. Fundamental Neuroscience. 4th ed. Oxford: Elsevier; 2012 [Chapters 13–15, 50]. Fig. 26.11, p. 591. With permission.

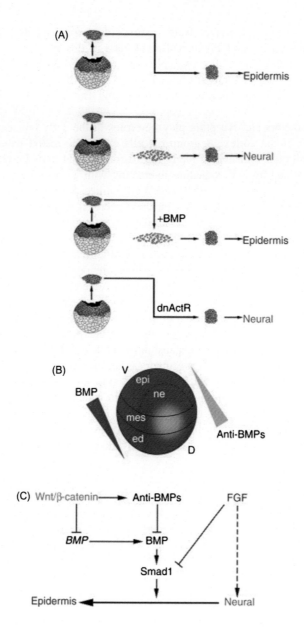

FIGURE 8.6

The BMP signaling pathway and its role in DV patterning of the ectoderm. (A) Experiments in Xenopus embryos that led to the default model: culture of animal cap explant results in epidermis differentiation; dissociation for several hours followed by reaggregation of animal cap tissue results in neural induction; the presence of BMPs during dissociation prevents neural induction and promotes epidermis formation; the expression of a dominant-negative Activin receptor results in neural induction even without dissociation. (B) Schematic model for DV patterning of the amphibian embryo. Spemann's organizer

(Continued)

the number of neurons within each column of the cortex without a marked change in the cortical surface area.

Very early in cortical development, cells containing fibrillary acidic protein (GFAP) and other astrocyte marker proteins derive from some of the NECs and become radial glial cells (RGCs). These cells are the earliest morphologically distinct cell in the primitive cortex. The RGC cell body is located in the VZ and layers directly above the VZ. These layers are named as the subventricular zone (SVZ) and the outer subventricular zone (OVZ). Only the RGCs found in the VZ contain a short process attached to the ventricle. However all RGCs have a long process that reaches the pial surface (Fig. 8.8).

The cortex thickens via the addition of an intermediate zone (IZ). The IZ contains the developing axons from the migrating neurons that will eventually form the white matter underlying the cellular layer or gray matter. The processes of the RGCs remain attached to the ventricular and pial surface respectively, making them ideal scaffolds for the migrating neurons from the SVZ (Fig. 8.8A). There is now accumulating evidence that there are molecular differences between human RGCs residing in the VZ and those in the OVZ. Most of the excitatory neurons are generated in the latter zone as well as the stem cells that sustain the proliferative niche.

As the neuronal precursors differentiate in the VZ and SVZ, one cell becomes a neuron while the other cell remains an RGC. However, cells destined to become neurons must avoid Delta/Notch signaling, which as described in Chapter 1, Classes of receptors, their signaling pathways, and their synthesis and transport, would cause them to become glia. They do so in part by expressing Numb, a cytosolic protein that antagonizes Notch signaling.

As shown in Fig. 8.8A, the neurons generated earliest in each column forming the deepest cortical layer, layer VI, migrate first. These neurons form a lamina, the axons of which project mainly to the thalamus. The thalamic nuclei are targets for sensory neurons including the retina. There are also projections to other regions of the cortex. Layer V neurons follow, migrating over layer VI neurons

---

◀ (*green*) on the dorsal (D) side of the embryo produces BMP antagonists that counteract the activity of BMPs (*blue*) on the ventral side (V). This results in a BMP activity gradient that patterns the DV axis of the embryo. Ventral ectoderm develops into epidermis (epi) under the influence of BMPs, whereas the neural plate (neurectoderm, ne) is induced on the dorsal side through BMP inhibition; mes, mesoderm; ed, endoderm. (C) The default model for neural induction including recent modifications: BMPs induce epidermal fate and inhibit neural induction via Smad1 activity; BMP inhibitors act as neural inducers by blocking BMPs; FGFs act as neural inducers by counteracting Smad1 and via BMP-independent mechanisms; Wnt/β-catenin signaling predisposes ectoderm for neural induction by both preventing the transcription of Bmp genes and inducing the expression of BMP inhibitors.

Squires LR, Berg D, Bloom FB, Du Lac S, Ghosh A, Spitzer NC. Fundamental Neuroscience. 4th ed. Oxford: Elsevier; 2012 [Chapters 13–15, 50]. Fig. 13.3, p. 290. With permission.

(A)

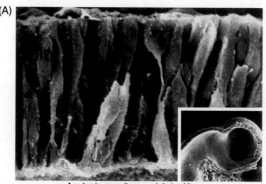

(B)

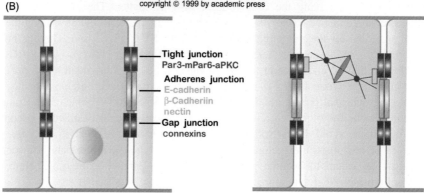

FIGURE 8.7

(A) The ventricular zone forms a pseudostratified columnar epithelium. Here neural progenitor cells have been visualized in the cerebral vesicle of a hamster embryo using scanning electron microscopy. Neuroepithelial cells are elongated bipolar cells that, at this early stage of development (E9.25), span the entire wall of the cerebrum. Some of the cells at the ventricular surface (bottom) appear spherical; these cells have retracted their cytoplasmic processes and are presumably rounding up in preparation for mitosis. Other rounded cells at the external surface (top) may be young neurons beginning to differentiate. (Inset) A low-power view of the hamster cerebral vesicle, corresponding roughly to that of a human embryo at the end of the first month of gestation. (B) Polarity proteins and vertebrate epithelia. Three classes of cell—cell junctions form between cells in a vertebrate epithelium. Conserved polarity proteins function in the formation of tight junctions (*red*). The cadherin and nectin cell adhesion proteins are required to form adherens junctions (*green*) and connexins form gap junctions (*purple*). During cell division in the neuroepithelium, the mPar6 polarity complex positions the spindle of the dividing cell.

Squires LR, Berg D, Bloom FB, Du Lac S, Ghosh A, Spitzer NC. Fundamental Neuroscience. 4th ed. Oxford: Elsevier; 2012 [Chapters 13–15, 50]. Fig. 15.9, p. 348. With permission.

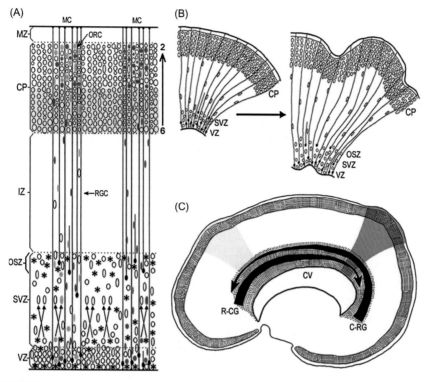

FIGURE 8.8

Radial unit model of the deployment of postmitotic migratory neurons and their settling pattern into the horizontal-laminar, inside-out, and vertical-columnar organization. (A) Neuronal progenitors in the proliferative ventricular and subventricular zones (VZ/SVZ/OSZ) and their progenies exhibit clonal heterogeneity (indicated by the *differently colored ellipses*). Several clones become intermixed in the SVZ, before migrating across the intermediate zone (IZ) along elongated shafts of the radial glial cells (RGC) into the cortical plate (CP). Newborn neurons bypass previously generated cells of the deeper layers (*yellow stripe*) in the inside-out sequence (layers VI to II) to participate in phenotypically and functionally heterogeneous mini-columns (MC) consisting of several ontogenetic radial columns (ORC). (B) Graphic explanation of the radial unit hypothesis of cortical expansion, by either preventing programmed cell death or increasing the rate of proliferation in mice, can produce a larger number of radial units that, constrained by the radial glial scaffolding, generate an expanded cellular sheet, which begins to buckle and transforms a lissencephalic (on the left) to the gyrencephalic (on the right) cerebrum. Based on studies in primates and experiments in mice. (C) Illustration of the concept, how opposing rostrocaudal (R-CG) and caudorostral (C-RG) molecular gradients, which form the protomap in embryonic VZ/SVZ lining cerebral ventricle (CV), can introduce new subtypes of neurons that migrate to the overlying CP and establish new cortical areas in the superjacent CP, indicated by yellow and orange color stripes based on experimental data in mice.

Geschwind DH, Rakic P. Cortical evolution: judge the brain by its cover. Neuron. 2013;80:633–647. Fig. 1.

With permission.

and producing axons that mainly project to subcerebral areas including the brainstem. Layer IV neurons migrate next and form short local axon projections within their own column. These neurons also are the main recipients of axons from the thalamus, which carry sensory information. Layer III neurons follow, producing axons that travel long distances within the same cerebral hemisphere, or to the other hemisphere via a large white matter tract, the corpus callosum. These axons form synapses with other excitatory neurons. Layer II neurons then migrate and project long axons as well. These reach either to other neurons in the same hemisphere or by way of the corpus callosum to neurons in the opposite hemisphere. Layer I, also called the molecular layer, is the last layer to be formed and is mainly composed of dendrites from neurons located in deeper layers. Also found in layer I are axons that are targeted to other cortical regions.

One theory explaining how the laminar structure is achieved during cortical development was proposed by Rakic and his colleagues. They postulate that all projection neurons come from a single class of RGC progenitor. At the earliest time in development a class of RGC is formed by cell division which gives rise to layer VI neurons (Fig. 8.8A). At later times, the RGCs sequentially produce layer V through II neurons as shown in Fig. 8.8A. Ultimately, the RGCs produce oligodendrocytes and astrocytes (Fig. 8.9). Thus the fate potential of a single RGC is progressively restricted over time.

A second theory that is gaining popularity is that there are a number of RGC subtypes, each giving rise to one or more projection neuron classes. Early in development, one RGC class gives rise to a group of layer VI neurons only. Different RGC subtypes give rise sequentially to layer V through II neurons. Finally, other RGCs give rise to oligodendrocytes and astrocytes sequentially during late development.

Excitatory neuronal migration depends on several proteins associated with microtubules. Lis1 and doublecortin are two gene products. It is not well understood how do they function. However, both proteins play important roles in nuclear migration from the ventricular to the pial surface. Mutations in either protein thicken the laminar layers by the addition of extra excitatory neurons. The laminar layers are also badly disrupted.

Another important signaling system involves Reelin, a glycoprotein that is secreted by the Cajal—Retzius neurons, which are found in the preplate and marginal zones on top of layer I and underlying the pia. The story of how Reelin was discovered is quite fascinating. In 1951, a mouse was characterized as having a reeling gait produced by poor motor coordination. Four decades later, the gene responsible for this condition was identified and the protein it coded was given the name, Reelin. Reelin binds to its receptors, ApoE receptor 2 and very-low-density lipoprotein receptor. After binding, the activated signal travels via cytoplasmic tyrosine kinases of the Src family. Ultimately signaling cascades switch on the adhesion protein cadherin, together with motility stimulating actin binding proteins, which together trigger motility. The signaling cascade ultimately results in the migrating neurons' detachment from their RGC scaffolds once they reach

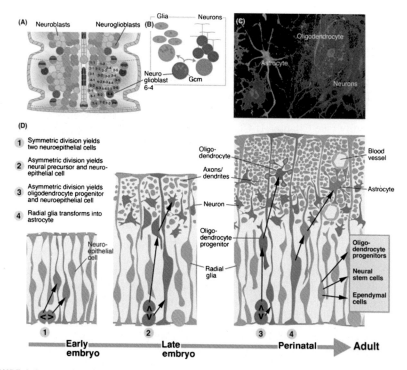

FIGURE 8.9

Specification of glial lineages. (A) Neuroblast map of the ventral nerve cord with glioblasts and neuroglioblasts indicated. Neuroblasts/neuroglioblasts of one hemineuromere are identified alphanumerically. (B) Neuroglioblast Nb6-4 expresses the glial regulatory protein Gmc (*green*). When this cell divides into two daughter cells, 6-4G and 6-4N, the Inscuteable complex and Miranda segregate Gmc into 6-4G, which thereby becomes specified as glioblast. 6-4N generates neurons. (C) Photograph of cell lineage obtained by injecting an individual neuroepithelial cell from mouse cortical progenitor cultured in a dish. Neural and glial-specific antibodies reveal the presence of both neurons (*purple*) and glia (astrocytes: *green*; oligodendrocytes: *blue*) in the lineage. (D) Schematic cross section of neural tube of early embryo (left), late embryo (middle), and around birth (right; see timeline at bottom). At early stage, symmetric divisions of neuroepithelial cells (1) result in an increased number of neural progenitors. At later stages, neuroepithelial cells (now synonymous with radial glial cells) start to divide asymmetrically, producing neurons (*red*), and maintaining their own number. Perinatally, production of neurons ceases, and neuroepithelial cells/radial glia switch to the production of oligodendrocyte progenitors (*blue*). Many radial glial cells delaminate and become astrocytes (*green*). Radial glia also give rise to the ependymal cells lining the ventricles of the adult CNS, to more oligodendrocyte progenitors, and to neural stem cells that remain active throughout adult life.

Squires LR, Berg D, Bloom FB, Du Lac S, Ghosh A, Spitzer NC. Fundamental Neuroscience. *4th ed. Oxford: Elsevier; 2012 [Chapters 13–15, 50]. Fig. 14.5, p. 317. With permission.*

their appropriate position. Mutations in Reelin produce an almost totally inverted laminar structure. There is evidence derived from animal studies that Reelin also plays a role in synaptogenesis.

Another group of receptor mediated trophic signaling systems has been identified in the developing neocortex. The Wnt, Frizzled signaling system described in Chapter 1, Classes of receptors, their signaling pathways, and their synthesis and transport, plays an important role. Another signaling receptor, a member of the integrin family (see Chapter 1: Classes of receptors, their signaling pathways, and their synthesis and transport), that interacts with extracellular matrix components is also important in cortical development. Finally two enzymes, Caspases 3 and 8, which play major roles in programmed cell death/apoptosis and will be discussed in Chapter 9, The importance of rapid eye movement sleep and other forms of sleep in selecting the appropriate neuronal circuitry; programmed cell death/apoptosis, are involved in cortical development. This is almost certainly due to the fact that approximately 50% of the neuronal precursor cells commit suicide at some stage of cortical development.

After maturation, the excitatory neurons contain a population of small, round Glu-containing synaptic vesicles (see Fig. 4.9). There is no convincing evidence that these vesicles contain other neurotransmitters. This is in direct contrast to the synaptic vesicles of GABAergic neurons, which will be discussed next.

The second major group of neurons that populate the mature cerebral cortex are the GABAergic inhibitory interneurons. They mainly form synapses locally on the pyramidal neurons located in the same column. While the pyramidal cells mainly have the same overall shape—a triangular cell body with apical, oblique, and basal dendrites and a single axonal arbor—interneurons are much more variable in shape, size, and dendrite number (see Fig. 9.4).

In contrast to the small, round Glu synaptic vesicles contained in mature excitatory neurons, the synaptic vesicles in GABAergic neurons have a small, clear, oval shape (see Fig. 4.9).

These presynaptic endings also contain larger, dense core vesicles, each having one or more of a wide variety of neuropeptides. Some of the more abundant of these are neuropeptide Y, vasointestinal peptide, somatostatin, and the calcium binding proteins, calretinin and parvalbumin.

The development and migration of the two major classes of cortical neurons also differ markedly. As mentioned above, pyramidal excitatory neurons are generated in the region of the cell layers surrounding the lateral ventricles, the VZ and the SVZ, and migrate in an inside-out manner to reach their final destinations. Pyramidal cells complete their migrations and send out axonal processes before birth.

In contrast, all of the GABAergic neurons are generated later than the excitatory neurons. The origin of the GABAergic interneurons is the lowest region of the forebrain. There are several nuclei located in this region termed the ganglionic eminences (GEs). There are five GEs in rodents: caudal, lateral, medial, preoptic, and septal (CGE, LGE, MGE, PGE, and SGE) respectively. In rodents, different portions of the GEs give rise to different neuropeptide-containing interneurons at different times. For example, somatostatin-containing neurons are generated in the medial GE at early times while vasointestinal peptidecontaining interneurons are generated at later times in the CGE.

In rodents, most of these GABAergic neuron classes form migrating streams that move tangentially into different regions and layers of the cortex. However, some remain in the striatum and other regions of the midbrain. Also cholinergic neurons of the striatum and other brain areas are generated in the MGE. However, primates' GABAergic interneurons arise from both the GEs and the cortical VZ and SVZ.

As mentioned previously, the GE-generated inhibitory neurons in both rodents and primates migrate tangentially into the cortex. During cortical development, the axons of the excitatory neurons reach the internal capsule, a large axon tract running to and from the cortex and thalamus respectively. When this occurs, the migrating GABAergic neurons encounter the axon tract. They then use very specific guidance cues involving trophic factor—receptor interactions to reach their final destinations in the laminae of the cortex.

Interestingly, even though the excitatory neurons migrate to form the laminae, the first postsynaptic elements to form on their dendrites are from the presynaptic regions of axons from the GABAergic neurons, which reach the cortical laminae later. These synapses initially are excitatory and only later become inhibitory. However, the GABAergic dendrites make synapses with each other.

The specific trophic factor—receptor interactions that regulate GABAergic neuron formation and migration are not well understood. However, one key transcription factor, DIX, controls many aspects of GABAergic cell migration and differentiation, including expression of the neuroligin—neurexin combination that is very important in synapse formation (see Fig. 4.16). Cells lacking DIX do not migrate properly or make appropriate synaptic connections.

RECOMMENDED BOOK CHAPTERS

Kandel ER, Schwartz JH, Jessel TM, Siegelbaum SA, Hudspeth AJ. *Principles of Neural Science*. 5th ed. New York: McGraw Hill; 2013 [Chapters 15, 17, 52, 53].

Squire LR, Berg D, Bloom FB, Du Lac S, Ghosh A, Spitzer NC. *Fundamental Neuroscience*. 4th ed. Oxford: Elsevier; 2013 [Chapters 13—15, 50].

Review Articles

Geschwind DH, Rakic P. Cortical evolution: judge the brain by its cover. *Neuron.*. 2013;80:633—647.

Lui JH, Hansen DV, Kriegstein AR. Development and evolution of the human neocortex. *Cell.*. 2011;146:18—36.

CHAPTER

The importance of sleep in selecting neuronal circuitry; programmed cell death/apoptosis

9

CHAPTER OUTLINE

9.1 RAPID EYE MOVEMENT SLEEP

Once a group of neurons in a particular brain region forms synapses with the various neurons they contact, the neurons that have formed appropriate connections (synapses) are selected for continued growth and survival. In contrast, the ones that are misconnected are destroyed by the process described below, that is, programmed cell death/apoptosis. The mechanism that the brain uses to select the neurons that have made the appropriate connections is called rapid eye movement (REM) sleep. It is a stage of sleep found only in mammals and birds. Also they are the only vertebrates to have neuronal programmed cell death/apoptosis during brain development, presumably because their developing brains contain so many neurons that many connectional errors are made. Therefore the neurons that make these misconnections must be eliminated (see Fig. 7.1). One can hypothesize that the evolution of REM sleep in mammals and birds was crucial to the creation of brains with a huge number of neurons compared with those in reptiles and amphibians.

During REM sleep, the period in which dreams mainly occur, the brain is not communicating with the spinal cord. Therefore no voluntary movements occur except for the eyes, which move rapidly from side to side. This REM led Dement to identify it as a unique period of sleep and hence, its name.

Once REM sleep has been activated in a given part of the brain, neurons fire very rapidly, releasing neurotransmitters and neuropeptides that are also capable of acting as trophic factors (Fig. 9.1). Neurons that have made the correct synaptic connections receive their trophic support and survive. In contrast, those that haven't made the appropriate connections, and thereby do not receive their trophic nutrients, are destroyed by activation of apoptosis. Fetuses are in REM sleep most of the time. Also the fetal period is when the greatest amount of neuronal

Receptors in the Evolution and Development of the Brain. DOI: https://doi.org/10.1016/B978-0-12-811012-6.00009-1

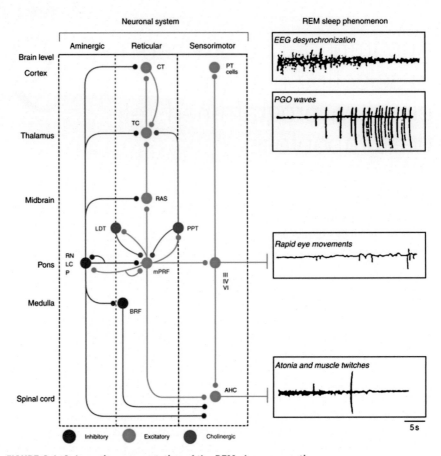

FIGURE 9.1 Schematic representation of the REM sleep generation process.

A distributed network involves cells at brain levels from the spinal cord to the cortex (listed on the left). The network is represented as a diagram of three neuronal systems (aminergic, reticular, and sensorimotor) that mediate REM sleep phenomena (depicted in tracings on the right). Postulated inhibitory connections are shown as red circles; postulated excitatory connections as green circles; and cholinergic pontine nuclei are shown as blue circles. It should be noted that the actual synaptic signs of many of the aminergic and reticular pathways remain to be demonstrated, and, in many cases, the neuronal architecture is known to be far more complex than indicated here (e.g., the thalamus and cortex). During REM, additive facilitatory effects on pontine REM-on cells are postulated to occur via disinhibition (resulting from the marked reduction in firing rate by aminergic neurons at REM sleep onset) and through excitation (resulting from mutually excitatory cholinergic–noncholinergic cell interactions within the pontine tegmentum). The net result is strong tonic and phasic activation of reticular and sensorimotor neurons in REM sleep. REM sleep phenomena are postulated to be mediated as follows: EEG

(*Continued*)

development occurs. The percentage of time spent in REM sleep declines with age, although some is present throughout life (Fig. 9.2). The necessity for trophic support, however, is postulated to last for a lifetime as will be discussed in detail in Chapter 18, Trophic factor—receptor interactions that mediate neuronal survival can in some cases last through life, and trophic deficits can produce connectional neurodegenerative diseases.

In some cases during development, neurons comprising whole circuits that are no longer being used stop firing, do not receive trophic support especially during REM sleep, and are therefore eliminated. An example of circuit elimination is the sucking reflex in infants. Once the neuronal circuitry essential for its performance is not necessary, the whole network of neurons involved in this reflex is eliminated.

The mechanism underlying REM sleep is as follows. Neurons in the brain stem and basal forebrain nucleus, which synapse on neurons in many parts of the brain, fire in anticipation of either waking or REM sleep. Also phasic activity of neurons in the pontine, geniculate nuclei, and occipital cortex correlate with REM sleep and with the phasic release of ACh. In contrast, norepinephrine and histamine containing neurons in the brain stem and hypothalamus, respectively, also synapse with neurons in many other regions of the brain, including the thalamus. These neurons are active during the four phases of non-REM or slow wave sleep. The circuits that they form are inactivated during REM sleep. In contrast to the tonic activity of thalamic relay neurons during the awake state, these neurons fire in bursts during slow wave sleep (Fig. 9.3). Neurons active during the four stages of slow wave sleep may also contribute to the selection of the appropriate wiring of the developing brain.

◀ desynchronization results from a net tonic increase in reticular, thalamocortical, and cortical neuronal firing rates. PGO waves are the result of tonic disinhibition and phasic excitation of burst cells in the lateral pontomesencephalic tegmentum. REMs are the consequence of phasic firing by reticular and vestibular cells; the latter (not shown) excite oculomotor neurons directly. Muscular atonia is the consequence of tonic postsynaptic inhibition of spinal anterior horn cells by the pontomedullary reticular formation. Muscle twitches occur when excitation by reticular and pyramidal tract motor neurons phasically overcome the tonic inhibition of the anterior horn cells. *III*, oculomotor; *IV*, trochlear; *V*, trigminial motor nuclei; *AHC*, anterior horn cell; *BIRF*, bulbospinal inhibitory reticular formation (e.g., gigantocellular tegmental field, parvocellular tegmental field, magnocellular tegmental field); *CT*, cortical; *LC*, locus coeruleus; *LDT*, laterodorsal tegmental nucleus; *mPRF*, meso- and mediopontine tegmentum (e.g., gigantocellular tegmental field, parvocellular tegmental field); *P*, peribrachial region; *PPT*, pedunculopontine tegmental nucleus; *PT cell*, pyramidal cell; *RAS*, midbrain reticular activating system; *REM*, rapid eye movement; *RN*, raphe nuclei; *TC*, thalamocortical.

Squire LR, Berg D, Bloom FB, Du Lac S, Ghosh A, Spitzer NC. Fundamental Neuroscience. 4th ed. Oxford: Elsevier; 2013 [Chapters 18, 40]. Fig. 40.9, p. 857. With permission.

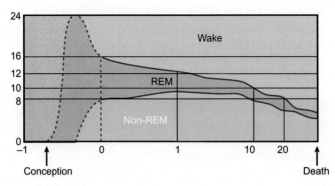

FIGURE 9.2 Portions of a 24-h day that are devoted to waking, rapid eye movement (REM) sleep, and non-REM (NREM) sleep change over a lifetime.

Although the timing of these changes in utero is not known with certainty (*dotted lines*), data from premature infants are consistent with REM sleep occupying most of life at a gestational age of 26 weeks. After 26 weeks, the time spent in waking increases until death.

Squire et al. 3rd ed. Fig. 42.3, p. 962. With permission.

Another group of neurons in the hypothalamus functions to maintain the waking state. These neurons release one of two similar neuropeptides, hypocretin 1 and 2, which bind to their receptors located on neurons in many parts of the brain. Patients with mutations in one of these proteins suffer from the condition narcolepsy and suddenly fall asleep during the day. This condition can also be caused by environmental agents.

Besides providing trophic support for the selection of the proper circuitry to be formed during brain development, sleep has several other important functions. It has recently been found that the cells of the brain shrink during sleep. This allows the easier removal of toxic molecules such as Aβ and glutamate that have accumulated during wakefulness. This removal of toxins occurs through channels that run alongside the vascular system and is called the glymphatic system. This system is analogous to the lymphatic system, which performs the same function in the periphery. The glymphatic channels are formed by the processes of astrocytes that lie next to the endothelial cells that comprise the brain's vascular system. The astrocytes contain a water transporter, AQP4, which pumps water molecules from the interstitial spaces in the brain into the glymphatic channels. These channels empty into the cerebrospinal fluid (CSF) and finally into the blood.

Most of the fluid in the brain is produced by ependymal cells, which comprise the choroid plexus. This group of cells is derived from the early neuroepithelium. The cells are columnar, with tight junctions on their luminal sides, preventing even small ions from reaching the brain (Fig. 9.4). The ependymal cells completely line the four ventricles of the brain, and each has a long flagellum

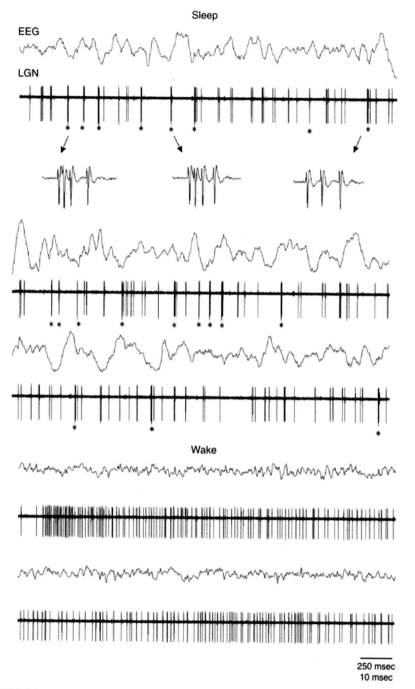

FIGURE 9.3

Two alternate firing modes of thalamic relay neurons: the bursting mode that is typical of slow wave sleep and the tonic (transmission) mode typical of waking. Traces show

(*Continued*)

that by beating moves liquid through the CSF. The ependymal cells each contain many molecules of the water transporter AQP1, which allow water to flow down its gradient from blood to brain. Also various ion transporters are present, among which are those that raise the K^+ concentration in the CSF compared with that in the blood (Fig. 9.5).

Another function of both slow wave and REM sleep is to consolidate and strengthen memories of events that occurred during the previous day. Regular memories are consolidated during slow wave sleep and emotionally charged memories during REM sleep. An example of the class of memories that are consolidated during non-REM sleep are memories that determine location in three-dimensional space, which involves the hippocampus. In contrast, memories of emotionally charged events occurring during the most recent period of wakefulness are recapitulated during REM sleep. This process involves both the hippocampus and the amygdala. In fact, when subjects are aroused during REM sleep, they remember emotionally charged events that occurred during the last period of wakefulness.

Finally, sleep is necessary for all multicellular animals that have been studied. One possible explanation is that animals have to remain at rest so they can replenish energy supplies that have been utilized during wakefulness. Also in choosing a safe resting place, animals can escape predators.

9.2 PROGRAMMED CELL DEATH/APOPTOSIS

Receptor—trophic factor signaling leading to the creation of a nervous system, formation of the cellular complex that produces myelinated axons, and the development of REM and slow wave sleep were all keys to the evolution of a big (in terms of cell number) brain. Another intricately regulated system that developed during evolution of multicellular organisms is the process of apoptosis, from the Greek word meaning "falling off"; as leaves do from a tree. An apoptosis-like

◀ recordings in the thalamic lateral geniculate nucleus (LGN) of the cat along with simultaneous EEG from the occipital cortex. The top 7 traces show burst firing during sleep and the resultant synchronized low-frequency, high-amplitude occipital EEG. The bottom 4 traces show recordings from these same sites during waking along with the resultant desynchronized, high-frequency EEG. Asterisks indicate bursting periods and the third trace shows an enlargement of the LGN channel during bursts. In the tonic mode during waking, firing frequency is proportional to the strength of a sensory input whereas during the bursting mode in sleep, transmission of sensory information to the cortex via the thalamus is blocked.

Squire LR, Berg D, Bloom FB, Du Lac S, Ghosh A, Spitzer NC. Fundamental Neuroscience. 4th ed. Oxford: Elsevier; 2013 [Chapters 18, 40]. Fig. 40.8, p. 856. With permission.

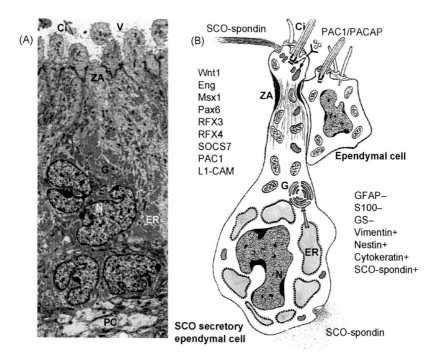

FIGURE 9.4 The secretory feature of the SCO ependymal cell.

(A) Electron micrograph of the secretory ependymal cells of the subcommissural organ in the adult mouse brain showing the well-developed cisternae of the endoplasmic reticulum (ER) filled with secretory material. The specific SCO secretion, called SCO-spondin, is secreted both apically into the third ventricular cavity (V) contributing to Reissner's fiber formation or basally contributing to the matrix of the posterior commissure (PC). (B) Scheme of a SCO secretory ependymal cell in comparison with a ventricular ependymal cell. Genes known to be involved in both SCO secretory cell identity and hydrocephalus are indicated on the left. They comprise developmental genes (Wnt1, Eng, Msx1, and Pax6), regulators of ciliary function (RFX3 and 4), a suppressor of cytokine signaling (SOCS7), a G protein-coupled receptor (PAC1), and a cell adhesion molecule (L1-CAM). The main features of the SCO secretory cells are indicated on the right. They do not express typical glial markers such as glial fibrillary acid protein (GFAP), glutamine synthetase (GS), and S100 protein but they contain vimentin, nestin, and cytokeratin intermediate filaments. The precursor form of the specific SCO secretion, SCO-spondin, is stored in the endoplasmique reticulum (ER) and modified in the Golgi apparatus (G) (*open arrow*) before being released both apically into the CSF and basally into the matrix of the posterior commissure. Downregulation of SCO secretion by altered PACAP/PAC1 signaling is associated to abnormal ciliary function (*double arrow*). *Ci*, cilium; *G*, Golgi apparatus; *N*, nucleus; *SCO*, subcommissural organ; *ZA*, zonulae adherens.

Meiniel A. Int J Biochem Cell Biol. *2007;39:463–8. Fig. 2. PMID:17150405. With permission.*

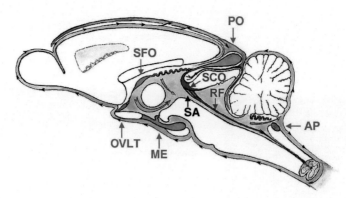

FIGURE 9.5 The subcommissural organ/Reissner's fiber complex and cerebrospinal fluid flow.

The subcommissural organ (SCO)/Reissner's fiber (RF) complex (*in red*) at the entrance of the sylvian aqueduct (SA) would facilitate the flow of the cerebrospinal fluid (*in orange*) through this narrow canal. The movement of the CSF flow is indicated by arrows. The CSF, secreted by the choroid plexus cells (*green waves*), circulates in a rostro-caudal direction from the lateral and third ventricles through the aqueduct of sylvius to the fourth ventricle and then into the subarachnoid space and central canal of the spinal cord. CSF is removed through the arachnoid villi into the venous circulation (*blue*). The other circumventricular organs are indicated in green. *AP*, area postrema; *ME*, median eminence; *OVLT*, organum vasculosum of the lamina terminalis; *PO*, pineal organ; *SFO*, subfornical organ.

Meiniel A. Int J Biochem Cell Biol. 2007;39:463–8. Fig. 1. PMID:17150405. With permission.

form of cell suicide evolved in prokaryotes as well and displays similar morphologic features to those found in multicellular organisms. These include cell condensation producing a darkened granular appearance and nuclear membrane condensation with chromosome destruction (Fig. 9.6). Small membrane-enclosed apoptotic bodies form that are engulfed by neighboring cells, mainly microglia, without the leakage of enzymes or other potentially toxic materials as is the case with necrosis. Another characteristic of apoptosis is that when the apoptotic cell's DNA is extracted and run on an acrylamide gel, one obtains a ladder-like pattern of fragments (Fig. 9.7).

Two other regulated cell death pathways share some molecular components with apoptosis, autophagy, and necroptosis. The former means literally "devouring oneself" and can occur during cellular starvation. It can be either a survival or death producing mechanism. The Nobel Prize in Physiology and Medicine was awarded in 2016 to Dr. Yoshinori Ohsumi for his elucidation of the molecular pathway that leads to autophagy in yeast. Autophagy in vertebrates has recently been shown to share a similar pathway as that in yeast.

Necroptosis usually occurs following a viral or bacterial infection or in the course of an inflammatory or neurodegenerative disease like Alzheimer's disease,

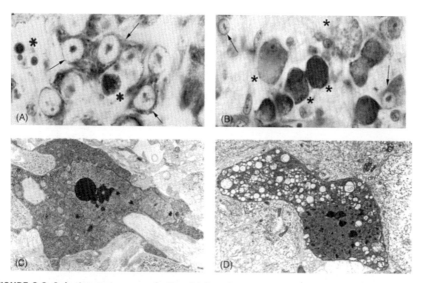

FIGURE 9.6 Spinal motor neurons in the chick embryo.

(A) Ventral horn from a control embryo. Note that only two cells (*asterisks*) are undergoing apoptotic PCD; the others appear normal (*arrows*). (B) Ventral horn from an embryo following treatment with an excitotoxin. Most neurons are undergoing a necrotic type of cell death (*asterisks*) whereas some appear normal (*arrows*). Apoptotic (C) and necrotic motor neurons (D) in the chick embryo spinal cord as seen with an electron microscope.

Squire LR, Berg D, Bloom FB, Du Lac S, Ghosh A, Spitzer NC. Fundamental Neuroscience. 4th ed. Oxford: Elsevier; 2013 [Chapters 18, 40]. Fig. 18.5, p. 411. With permission.

which will be discussed in Chapter 18, Trophic factor–receptor interactions that mediate neuronal survival can in some cases last through life, and trophic deficits can produce connectional neurodegenerative diseases. Neither autophagy nor necroptosis play important roles in the programmed cell death that occurs during neural development.

As was emphasized in Chapter 7, A testable theory to explain how the wiring of the brain occurs, apoptosis/programmed cell death is critical for the development of the big brain characteristic of birds and mammals. Trophic factor deprivation in neurons that fail to make appropriate connections trigger this pathway. Studies in Dr. Robert Horwitz's laboratory that led to his receiving the Nobel Prize in Physiology and Medicine in 2002 helped to elucidate the molecular mechanism of apoptosis. Genetic analysis of the round worm *Caenorhabditis elegans* revealed three genes that specifically function in a system that controls neuronal cell death during development. These are called CED-9, CED-4, and CED-3, which correspond to Bcl-2 like, Apaf-1 like, and Caspase-8 like in vertebrates. Several studies soon reported homologues of these genes in all classes of multicellular organisms including mammals and insects. These gene products correspond to three different families of proteins. One class is the adapter protein,

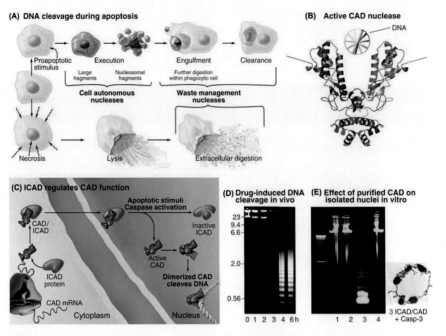

FIGURE 9.7 Nucleases digest the cellular DNA during apoptosis.

(A) In apoptosis, the DNA is digested first to large fragments and later to nucleosome-sized pieces by cell autonomous nucleases expressed within the dying cell. Waste management nucleases made by other cells also have an essential role in cleaning up apoptotic and necrotic debris. (B) The predominant cell autonomous nuclease (CAD) has a scissors-like structure. (C) ICAD is an inhibitory chaperone for CAD, promoting its folding on the ribosome and continuing as an inhibitor when CAD is stored in the nucleus. ICAD cleavage leads to CAD activation. (D) Cleavage of the chromosomal DNA by CAD during chemotherapy-induced apoptosis of a leukemia cell line. DNA separated according to size by electrophoresis on an agarose gel was stained with ethidium bromide. The ladder of bands reflects cleavage between adjacent nucleosomes. (E) Activated CAD causes chromatin condensation and appearance of an apoptotic morphology in isolated cell nuclei. Cloned CAD and ICAD were expressed together in *Escherichia coli* and incubated with nuclei. ICAD cleavage with caspase-3 released active CAD, which degrades the nuclear DNA (lane 3). Other lanes: DNA gel size markers (left); nuclei incubated with buffer or caspase-3 alone (lanes 1 and 2, respectively); same experiment as in lane 3 but performed by using a mutant ICAD that could not be cleaved by caspase-3 (lane 4). To the right is an electron micrograph of a thin section of one nucleus with condensed chromatin at the nuclear periphery. *CAD*, caspase-activated deoxyribonuclease; *ICAD*, inhibitor of caspase-activated deoxyribonuclease; *mRNA*, messenger RNA.

Pollard TD, Earnshaw WC, Lippincott-Schwartz J, Johnson GT. Cell Biology. 3rd ed. Philadelphia, PA: Elsevier; 2017. Fig. 46.13, p. 807. With permission.

Apaf-1, that is necessary to activate a cascade of cysteine proteases called caspases. This second class is composed of proteases that produce a sequential cascade, which degrades many proteins in a cell destined to self-destruct. The third class of proteins consists of members of a large family that regulates apoptosis, either positively or negatively. This class, called the Bcl-2 family, contains a specific domain, the BH domain. These proteins bind to mitochondria and either suppress or activate the death signal. Fig. 9.8 describes the apoptotic pathway in more detail.

There are two regulated pathways in neurons that lead to apoptosis. One is termed the intrinsic pathway and involves proteins or other factors in the cell that produce mitochondrial damage. Damage to the mitochondria impairs the synthesis of ATP, the energy-providing molecule of the cell. This in turn leads to the mitochondrial membrane becoming permeable to relatively large molecules including the small protein cytochrome C, which is an essential participant in the production of ATP, leading to a cascade of processes that produce apoptosis (Fig. 9.8).

The other pathway, the extrinsic pathway, is often triggered by the absence of trophic factor binding to its receptor. One protein, the absence of which triggers apoptosis, is the neurotrophin nerve growth factor (NGF), which was the first trophic factor to be isolated. Dr. Rita Levi-Montalcini and Dr. Stanley Cohen won the Nobel Prize in Physiology and Medicine in 1986 for this discovery.

NGF or another neurotrophin family member normally binds to its receptor, one of the Trk tyrosine kinase receptor family discussed in Chapter 1, Classes of receptors, their signaling pathways, and their synthesis and transport. This binding occurs either in the presence or absence of the low affinity neurotrophin receptor p75. Then the Trk receptor dimerizes, and the two subunits phosphorylate one another. This in turn activates protein kinase B (PKB) via phosphoinoside-3 kinase. PKB then phosphorylates Bad, a member of the Bcl-2 family, which are either pro- or antiapoptotic. When Bad is phosphorylated it is inactivated, thus allowing the cell to survive. On the other hand, if a neuron does not receive its trophic factor support, Bad is dephosphorylated and binds to Bcl-2, an antiapoptotic member of the family, which in turn causes the proapoptotic family members, Bak and Bax, to oligomerize. This oligomer in turn forms a mitochondrial pore that allows cytochrome C to escape from the mitochondria and binds Apaf-1. This binding causes the conversion of procaspase-9 to its active form leading to the proteolysis of many cellular proteins by other caspases, which are converted by caspase 9 to their active forms (Fig. 9.8). Activated DNAases and RNAases degrade the nucleic acids.

Another trigger for the extrinsic apoptosis pathway is the low affinity neurotrophin receptor p75. p75 binds to Trk A, B, or C in the presence of the appropriate neurotrophin and serves as a trophic signal. However, in the absence of a supply of a neurotrophin, p75 becomes a "death receptor" triggering an irreversible process leading to apoptotic cell death. Another death receptor is the Fas receptor, which triggers the same death pathway as does p75 (Fig. 9.9). A key feature of both pathways is the prevention of the leakage of damaging cellular

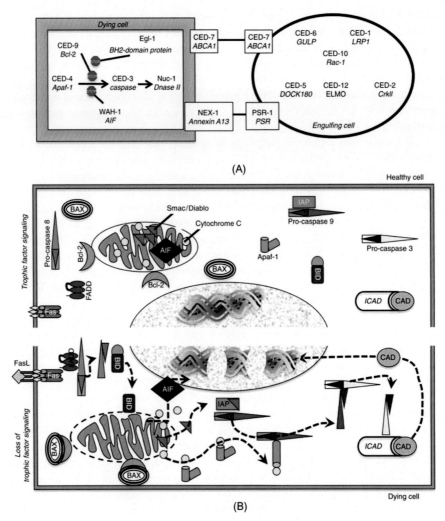

FIGURE 9.8 (A) Schematic representation of the major steps in the developmental PCD pathway of neurons in the nematode worm Caenorhabditis *elegans*.

Both fly and mammalian homologues have been identified supporting the hypothesis that mechanisms of PCD are evolutionarily conserved. The mammalian homologues are listed in italics below the name of the *C. elegans* gene. (B) Some of the major events involved in mammalian PCD are illustrated. Healthy cells (top) receive signals to survive or lack signals to die (on left). Cell death mediators are present in healthy cells, but their location and/or association with regulators prevent activation of the cell death pathways. In cells that receive appropriate signals to die (bottom), cell death−specific pathways are activated. These pathways involve a permeabilization of the mitochondria membrane and release of factors that either directly or indirectly activate cell death−specific events. In

(*Continued*)

proteases, DNAases, RNAases, and other toxic substances into the exterior milieu. During apoptosis, the cell is broken up into small membrane vesicles called apoptotic bodies that each contain small amounts of cellular debris. On the surface of each apoptotic body are many molecules of the phospholipid phosphatidyl serine, which under normal conditions is only found in the inner leaflet of the plasma membrane. However, when apoptosis is triggered, phosphatidyl serine migrates to the outer leaflet and is recognized by its receptors on microglial cells. The microglia then bind to and phagocytose the apoptotic bodies and digest their contents in lysosomes.

Two time frames of nerve cell development can be distinguished with respect to the occurrence of a large number of apoptotic events. The first involves early neural precursor cells, and is likely triggered by inaccessibility of some early neural precursors to a class of early cell division and differentiation proteins including members of the Wnt, Hedgehog, fibroblast growth factor, and BMP families. These can also be considered to be trophic factors that bind to specific receptors as was discussed in Chapter 1, Classes of receptors, their signaling pathways, and their synthesis and transport.

The second time frame during which apoptosis occurs follows the maturation of neurons that have migrated, put out processes, and made connections with other neurons. This event is triggered by the lack of crucial trophic factors including members of the neurotrophin family as described previously. Many other neuropeptides and classical neurotransmitters also function as trophic factors during brain development, as has been discussed previously, and their lack can also trigger apoptosis.

As emphasized in Chapter 14, The roles of receptors and trophic factors in the development of the enteric nervous system and in the connections between viruses, the microbiota, ENT, and brain, apoptosis is essential in mammals and

◄ neurons that fail to obtain NTF support (bottom), the proapoptotic gene Bax interacts with and inhibits the antiapoptotic gene Bcl-2 in mitochondria. This results in the release from mitochondria of cytochrome-c, which forms a complex with Apaf-1 and caspase-9 that in turn activates downstream caspases such as caspase-3 that ultimately directly or indirectly degrade diverse nuclear and cytoplasmic targets. These degradative changes are what define apoptosis and result in eventual engulfment and phagocytosis of the apoptotic cell. In some situations, two additional molecules released from mitochondria (along with cytochrome-c) are the proapoptotic proteins AIF, which can degrade the nucleus independent of caspases, and Smac/Diablo, which can inhibit IAP and promote the apoptotic pathway via caspase-9 and caspase-3. Although not shown here, in some situations developing neurons undergoing PCD activate cell cycle proteins that also serve a signaling function required for apoptosis. It is critical to note that all neurons do not use the identical pathways for cell death. For example, cell cycle proteins can mediate the death of sensory neurons; however, they do not play a role in motor neuron PCD.

Squire LR, Berg D, Bloom FB, Du Lac S, Ghosh A, Spitzer NC. Fundamental Neuroscience. 4th ed. Oxford: Elsevier; 2013 [Chapters 18, 40]. Fig. 18.7, p. 414. With permission.

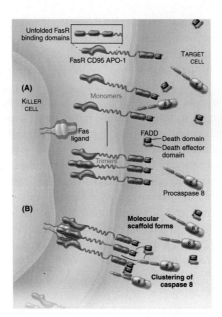
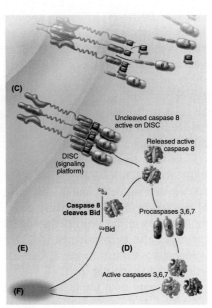

FIGURE 9.9 Extrinsic cell death pathway.

The pathways shown are downstream of the Fas cell death receptor. (A) Preligation. (B) Ligand docked on a trimerized receptor. (C) Release of active caspase-8. (D) Activation of effector caspases. (E) Activation of the intrinsic death pathway (see Fig. 46.17). (F) Death. *DISC*, death-inducing signaling complex; *FADD*, Fas-associated death domain.

Pollard TD, Earnshaw WC, Lippincott-Schwartz J, Johnson GT. Cell Biology. 3rd ed. Philadelphia, PA: Elsevier; 2017. Fig. 46.18, p. 811. With permission.

birds to select the proper brain circuitry. As was also discussed in Chapter 7, A testable theory to explain how the wiring of the brain occurs, this complex and demanding process is analogous to the following: If there are enough monkeys and enough typewriters, at least one monkey will certainly type *Hamlet*. The difficulty is selecting which monkey and which typewriter has succeeded. During the assembly and wiring of the incredibly complex human brain, apoptosis plays a crucial role in this selection process.

As was discussed in Chapter 7, A testable theory to explain how the wiring of the brain occurs, apoptosis in the developing nervous system does not appear to play a key role until the evolution of mammals and birds. Species with smaller brains contain fewer neural cells. Therefore no selection is required to establish the correct circuitry. However, in both birds and mammals with a great number of neurons and glia, selection is required to guarantee the formation of the circuits necessary to appropriately carry out the myriad functions that the brain is required to perform. In support of this argument, a number of studies indicate that between 30% and 50% of the neurons and glial precursors in the developing brains of mammals and birds are eliminated by apoptosis at some stage of maturation.

GENERAL REFERENCES

Recommended Book Chapters

Kandel ER, Schwartz JH, Jessel TM, Siegelbaum SA, Hudspeth AJ. *Principles of Neural Science*. 5th ed. New York: McGraw Hill; 2013 [Chapters 51, 53].

Squire LR, Berg D, Bloom FB, Du Lac S, Ghosh A, Spitzer NC. *Fundamental Neuroscience*. 5th ed. Oxford, UK: Elsevier; 2013 [Chapters 18, 40].

Review Articles

Ackermann S1, Rasch B. Differential effects of non-REM and REM sleep on memory consolidation?. *Curr Neurol Neurosci Rep*. 2014;14:430.

Hobson JA, Friston KJ. Waking and dreaming consciousness: neurobiological and functional considerations. *Prog Neurobiol*. 2012;98:82−98.

Horvitz HR. Worms, life, and death (Nobel lecture). *Chembiochem*. 2003;4:697−711.

Iliff JJ, Wang M, Liao Y, et al. A paravascular pathway facilitates CSF flow through the brain parenchyma and the clearance of interstitial solutes, including amyloid β. *Sci Transl Med*. 2012;4. 147ra111.

BDNF and endocannabinoids in brain development: neuronal commitment, migration, and synaptogenesis

10

CHAPTER OUTLINE

10.1 BRAIN DERIVED GROWTH FACTOR

Brain derived growth factor (BDNF) is one of the four-member neurotrophin family that also includes nerve growth factor and neurotrophin 3 and 4. Only BDNF will be discussed here because it plays an extremely important role in brain development. The other family members are essential for the growth of the peripheral nervous system but play minor roles in brain development.

BDNF is synthesized in the endoplasmic reticulum, as pre-pro-BDNF, and then is cleaved to the pro species in the Golgi apparatus. After conversion to the pro form, It is transported to axonal and dendritic endings in both constitutive and regulated secretory vesicles. The final conversion to mature BDNF occurs in these vesicles or after secretion. Resolving this question is important since pro-BDNF does not bind the high-affinity BDNF receptor, Trk B. Instead it binds to the low-affinity neurotrophin receptor, p75, which as discussed in Chapter 9, The importance of rapid eye movement sleep and other forms of sleep in selecting the appropriate neuronal circuitry; programmed cell death/apoptosis, potentially activates apoptosis (Fig. 10.1).

When BDNF acts through the TrkB receptor, dimerization and subsequent phosphorylation of the dimeric receptor set off a signaling cascade. This cascade can result in many enzymatic processes that play key roles at the various stages of brain development concluding with the successful wiring of the brain circuitry. While the BDNF-TrkB signaling pathway is thought to begin at the plasma membrane through stimulation of a variety of signaling pathways, there are also

Receptors in the Evolution and Development of the Brain. DOI: https://doi.org/10.1016/B978-0-12-811012-6.00010-8

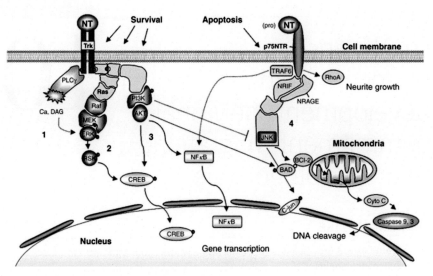

FIGURE 10.1 Trk and p75NTR signaling pathways.

Neurotrophin (NT) dimers bind two Trk receptor monomers or one p75NTR to activate survival (1–3) or apoptotic (4) signaling pathways. Ligand binding initiates Trk receptor transduction by the phosphorylation (indicated by *small red dot*) in the cytoplasmic domains. The activated Trk kinase docks adaptor and linker proteins (*light blue*), which engage signaling cascades often containing multiple kinases (*red*). The three principal signaling pathways illustrated are the (1) phospholipase C pathway, (2) Ras-MAP kinase pathway, and (3) PI-3 kinase pathway. These pathways lead to nuclear translocation of transcription factors (*green*), such as CREB and NFκB, and ultimately regulation of gene expression. The binding of neurotrophins or proneurotrophins to p75NTR can activate BAD via the JNK cascade and eventually engages, via release of effectors from mitochondria, caspases involved in apoptosis. There are several venues for cross-talk between the survival and proapoptotic signaling cascades as indicated. *BAD*, Bcl2-antagonist of cell death; *Bcl2*, B-cell lymphoma 2; *Ca*, calcium; *CREB*, cyclic AMP response element binding protein; *Cyto C*, cytochrome C; *DAG*, diacylglycerol; *JNK*, C-Jun N-terminal kinases; *NFκB*, nuclear factor-kappa B; *NRAGE*, neurotrophin receptor interacting MAGE (melanoma antigen gene expression) homolog; *NRIF*, neurotrophin receptor interacting factor; *p75NTR*, p75 neurotrophin receptor; *(pro)NT*, proneurotrophin; *Raf, Ras, and RhoA*, small GTPases; *RSK*, ribosomal S6 kinase.

data indicating that the BDNF-Trk complex can be internalized and become incorporated into signaling endosomes that can be active for long distances (Fig. 10.1). For example the binding of BDNF to TrkB at the axonal ending during synaptogenesis can trigger the internalization and retrograde transport of active signaling endosomes to the cell nucleus, eventually leading to changes in gene transcription. There is also evidence that neurotrophins and other trophic factors can be secreted from large peptide containing vesicles at the postsynaptic ending. Receptors enter the nucleus and turn on genes essential for survival. Growth cones of both elongated axons and dendrites also secrete BDNF and other trophic factors and contain receptors recognizing these trophic molecules.

There also appears to be a major difference in the morphological changes BDNF induces that is determined by the region of the cell surface in which BDNF/Trk B binding occurs. An example of this difference is the finding that when BDNF is infused into the superior colliculus, a target of retinal ganglion cell (RGC) axons, dendritic growth occurs in RGCs in the retina. In contrast when BDNF is added to the vitreous, which lies directly above the RGC layer of the retina, inhibition of dendritic growth occurs.

Even before overt neural differentiation occurs, BDNF plays an important role in the regulation of neuronal stem cell survival and differentiation. Trk B receptors are present on stem cells, and there is evidence indicating that treatment of these cells for 3 days promotes their differentiation into neurons at week 2. After 21 days in culture, BDNF no longer increases neuronal survival and differentiation, instead acting to promote axonal and dendritic growth. In contrast, NGF and NT-3 and their respective receptors, Trk A and C, which have limited distribution on neurons in the CNS, can serve as survival factors for these neurons.

Exposure of cultured hippocampal neurons to BDNF also promotes the conversion of one process in every neuron into an axon. It appears that autocrine secretion of BDNF by the neuron may be responsible, since specifically blocking BDNF synthesis also prevents axon formation. Another example of the importance of autocrine BDNF secretion is a very recent study indicating that RGCs produce their own BDNF and Trk B receptors until they reach their target territories. Then the neurons cease to produce BDNF and become dependent on target-derived BDNF.

There is numerous in vitro evidence demonstrating that BDNF and other neurotrophins can act as tropic factors for specific neuronal subtypes. In contrast, in vivo studies employing mice in which the BDNF gene has been deleted after the time when the mice require BDNF to survive show that CNS neuronal processes can still find their way to their target territories. However, these processes do not form correct synaptic connections. Numerous studies involving cerebral cortical, hippocampal, thalamocortical, and RGCs have demonstrated this fact.

However, during early neuron development, growth cones appear to make transient contacts with one another and then withdraw the connections and move in a different direction. These changes in the direction of movement, at least in vitro, depend on a continual supply of BDNF.

As mentioned previously, there is rapidly accumulating data indicating that BDNF plays a crucial role in synapse formation and maintenance. Synapses have been shown to be dynamic structures both during development and afterward. To summarize a great deal of data in many mammalian species, the dictum "use it or lose it" seems to generally apply. Synapses that are actively firing appear to be stabilized, and contain more synaptic vesicles in the presynaptic endings and enlarged postsynaptic endings. The opposite is true of inactive synapses, leading in extreme cases to the total disappearance of the synapse. Even though these synaptic changes are not as dramatic as the apoptosis of neurons that do not make their appropriate synaptic and trophic connections, they still play a key role in determining the wiring of the brain during its development. After the brain matures, when programmed cell death becomes of less importance, synaptic changes, also referred to as synaptic plasticity, become of critical importance. This topic will be discussed in greater detail in Chapter 15, Synaptic pruning during development and throughout life depends on trophic factor interactions: the corollary described in Chapter 14, postulating the quantization of trophic factor requirements, provides a plausible explanation for the survival of neurons partially deprived of their full complement of trophic factors.

The importance of BDNF in the refinement of synaptic circuitry in the developing brain is demonstrated by its role in the development of ocular dominance columns in the primary visual cortex. Thalamocortical neurons projecting to layer 4 pyramidal neurons segregate to form eye-specific columns. This segregation is caused by the noncoincident neuronal activities in the two eyes. There is a critical period during which the formation of visual dominance columns occurs. If an animal is deprived of vision in one eye during this period, these columns do not form (Fig. 10.2). Hubel and Wiesel first described this process, for which they were awarded the 1981 Nobel Prize in Physiology and Medicine.

(A)   Normal   (B)   Deprived eye   (C)   Nondeprived eye

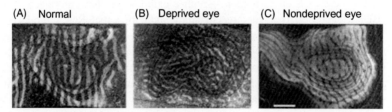

FIGURE 10.2 Critical period plasticity of the anatomical projections representing the left and right eyes in the visual cortex of monkeys.

These are dark field autoradiographs of tangential sections through layer 4 of the primary visual cortex, showing the pattern of transsynaptic labeling (bright areas) from the thalamic lateral geniculate nucleus (LGN) that resulted from an injection of a radioactive tracer into one eye; the intervening dark areas represent LGN input from the opposite (noninjected) eye. Scale bar indicates 2.5 mm. (A) The pattern of LGN input to the visual cortex in a normal monkey. (B) Severely reduced LGN input from the deprived eye of a monkey that experienced monocular deprivation from 3 weeks until 7 months old. (C) Greatly expanded LGN input from the nondeprived eye of a monkey that experienced monocular deprivation from 2 to 18 months of age.

Squire LR, Berg D, Bloom FB, Du Lac S, Ghosh A, Spitzer NC. Fundamental Neuroscience. *4th ed. Oxford: Elsevier; 2013 [Chapters 18, 21, 44]. Fig. 21.7, p. 487. With permission.*

BDNF plays a key role in establishing the optical dominance columns. Infusion of BDNF antibodies into the brain or preventing TrkB signaling with the injection of TrkB-specific antibodies blocks ocular dominance from occurring. It is now established that BDNF is important for the development of GABAergic synapses in the visual cortex, and transgenic mice overexpressing BDNF show an early maturation of these synapses. The enhanced development of GABAergic synapses in turn results in a decreased period for the establishment of ocular dominance columns.

This example and many others have led to the proposal that BDNF may be a key factor in the continuous modulation of synaptic circuitry that occurs during and after development and throughout life. One major process in which BDNF plays a critical role is long-term potentiation (LTP). LTP is thought to be a critical event in memory formation. However, before the role of BDNF in LTP is considered, it is necessary to explain the mechanisms underlying this process.

LTP was first described by Bliss and Loma using rats, in three different axonal pathways in the excitatory neurons of the hippocampus. If one gives a brief high-frequency train of electrical impulses to an axonal pathway, it leads to an increased amplitude of the excitatory postsynaptic potential in the targeted neurons (Figs. 10.3 and 10.4).

In one pathway, the commissural pathway, the postsynaptic neurons are stimulated by the activation of a class of glutamate receptors, the NMDA receptors. As described in Chapter 1, Classes of receptors, their signaling pathways, and their synthesis and transport, these receptors normally have a Mg^{++} ion in the ion channel, allowing the influx of $Na+$ but blocking Ca^{++} ions from entering the cell. The high-frequency stimulation, however, produces a change in the conformation of the NMDA receptor, causing Mg^{++} to leave the channel. This in turn allows a large amount of Ca^{++} to enter the postsynaptic ending, activating Ca^{++} dependent kinases. These kinases phosphorylate proteins, which ultimately produce a prolonged sensitivity to glutamate release from the presynaptic cell. Also retrograde messenger molecules are released from the postsynaptic cell that activate the presynaptic cell. This prolonged sensitivity can last 1−4 hours and is known as the early phase of LTP (Fig. 10.5).

If instead of giving one high-frequency stimulation four or more are given, another type of LTP is induced that can last for 24 hours or more. This type is called late phase LTP. Several aspects of this phase are of great importance. One is that it requires new protein synthesis and can be blocked by inhibitors of protein synthesis. These same inhibitors also block memory consolidation, which shows the same time course and in many cases involves the hippocampus.

A remarkable function that connects LTP to memory formation is the behavior of place cells in the hippocampus. Each of these cells by its firing pattern provides the rat with its precise location in three-dimensional space (Fig. 10.6). The discovery of place cells resulted in a Nobel Prize in Physiology and Medicine being awarded in 2014 to their discoverer, Dr. John O'Keefe. The other cell class that is involved in the three-dimensional spatial orientation of the rat is the

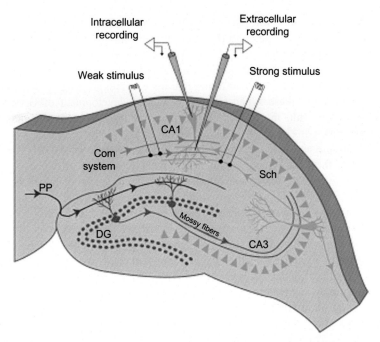

FIGURE 10.3 Schematic of a transverse hippocampal brain slice preparation from the rat.

Two extracellular stimulating electrodes are used to activate two nonoverlapping inputs to pyramidal neurons of the CA1 region of the hippocampus. By suitably adjusting the current intensity delivered to the stimulating electrodes, different numbers of Schaffer collateral/commissural (Sch/com) axons can be activated. In this way, one stimulating electrode was made to produce a weak postsynaptic response and the other to produce a strong postsynaptic response. Also illustrated is an extracellular recording electrode placed in the stratum radiatum (the projection zone of the Sch/com inputs) and an intracellular recording electrode in the stratum pyramidal (the cell body layer). Also indicated is the mossy fiber projection from granule cells of the dentate gyrus (DG) to the pyramidal neurons of the CA3 region.

Squire LR, Berg D, Bloom FB, Du Lac S, Ghosh A, Spitzer NC. Fundamental Neuroscience. 4th ed. Oxford: Elsevier; 2013 [Chapters 18, 21, 44]. Fig. 47.6, p. 1017. With permission.

entorhinal grid cell. The entorhinal cortex directly innervates the hippocampus and in turn receives input from the hippocampus (Fig. 10.6). The discoverers of grid cells, Edward Moser and May-Britt Moser, shared the 2014 Nobel Prize with Dr. O'Keefe.

There is now strong evidence that secreted BDNF plays a critical role in the activation of LTP. BDNF can function by activating both the pre- and postsynaptic neuron. In early LTP it functions by inhibiting the release of GABA onto the dendritic spines, thereby stimulating the firing of the Glu neurons and inducing early LTP. BDNF, probably secreted by the presynaptic terminals, also contributes to the

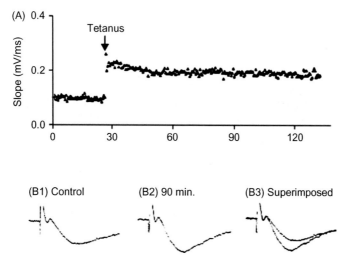

FIGURE 10.4 LTP at the CA3—CA1 synapse in the hippocampus.

(A) Test stimuli are delivered repeatedly once every 10 s while the strength of the synaptic connection is monitored. Strength can be assessed by the amplitude of the extracellularly recorded EPSP or, as was done in this example, as the slope of the rising phase of the EPSP, which provides an accurate reflection of its strength. To induce LTP, two 1 s, 100 Hz tetani were delivered with a 20-s interval. Subsequent test stimuli produce enhanced EPSPs. The enhancement is stable and persists for at least 2 h. Examples of extracellularly recorded field EPSPs before (B1) and 90 min after the induction of LTP (B2). In B3 the traces from B1 and B2 are superimposed.

Squire LR, Berg D, Bloom FB, Du Lac S, Ghosh A, Spitzer NC. Fundamental Neuroscience. 4th ed. Oxford: Elsevier; 2013 [Chapters 18, 21, 44]. Fig. 47.7, p. 1017. With permission.

formation of new synapses necessary for late LTP to occur concomitantly with memory formation. This BDNF function occurs via the stimulation of new protein synthesis and through the rearrangement of the actin cytoskeleton (Fig. 10.7). Finally, there is evidence suggesting that BDNF serves as the major activating molecule for presynaptic neurotransmitter release. This topic will be discussed in greater detail next when we describe the role of endocannabinoids in modulating synaptic activity.

BDNF is also necessary for the development of LTP. Prenatal stress, which inhibits LTP formation, also decreases the level of BDNF in the brain.

10.2 ENDOCANNABINOIDS

Marijuana's active ingredients include the hallucinogenic delta 9 tetrahydrocannabinol and a number of others. Marijuana has been used as a psychoactive agent

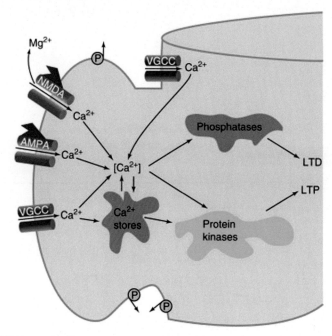

FIGURE 10.5 Events leading to some forms of LTP and LTD.

The schematic depicts a postsynaptic spine with various sources of Ca^{++}. The NMDA receptor channel complex admits Ca^{++} only after depolarization removes the Mg^{++} block in the presence of bound glutamate. Calcium may also enter through the ligand-gated AMPA receptor channel or voltage-gated calcium channels (VGCC), which may be located on the spine head or dendritic shaft. Calcium pumps (P), located on the spine head, neck, and dendritic shaft, are hypothesized to help isolate Ca^{++} concentration changes in the spine head from those in the dendritic shaft.

Squire LR, Berg D, Bloom FB, Du Lac S, Ghosh A, Spitzer NC. Fundamental Neuroscience. 4th ed. Oxford: Elsevier; 2013 [Chapters 18, 21, 44]. Fig. 47.9, p. 1019. With permission.

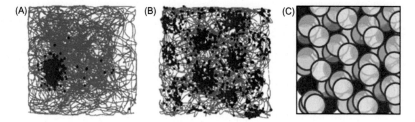

FIGURE 10.6 An example of a place cell (A) and a grid cell (B) recorded from a rat hippocampal and entorhinal neuron, respectively.

The gray lines show the path of a rat as it explored a square recording chamber, and the red dots show the locations along that path at which an example hippocampal (A) or entorhinal (B) neuron fired action potentials. The right panel (C) illustrates activity from multiple grid cells, and one idea is that place fields might relate to points of overlap from a collection of grid fields.

Squire LR, Berg D, Bloom FB, Du Lac S, Ghosh A, Spitzer NC. Fundamental Neuroscience. 4th ed. Oxford: Elsevier; 2013 [Chapters 18, 21, 44]. Fig. 48.8, p. 1037. With permission.

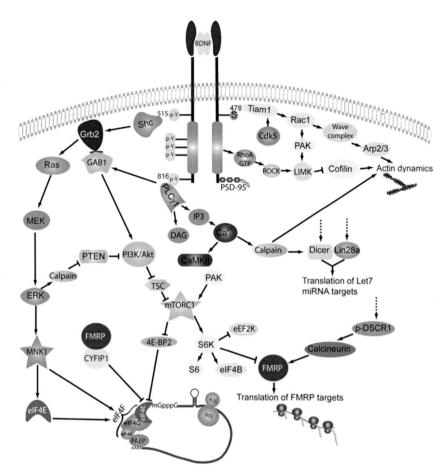

FIGURE 10.7 TrkB receptor coupling to translational control and actin cytoskeletal dynamics in dendritic spines.

This illustration depicts the majority of the signaling pathways and mechanisms discussed in this review of relevance for functional and structural plasticity in L-LTP. The summary is based on data obtained in different brain regions, experimental preparations, and paradigms. Many of the downstream pathways have not yet been shown to be regulated by endogenous BDNF in the context of L-LTP. Current evidence suggests that TrkB regulates multiple forms of mRNA-specific translation, controlled by CYFIP1, FMRP, and microRNAs, as well as more general regulation of translation initiation through mTORC1 and MNK. Dual regulation of translation and cytoskeletal dynamics by BDNF is another emerging theme. Reciprocal mechanisms may exist whereby cytoskeletal regulation promotes local translation, and translated proteins (such as Arc and LIMK1) influence actin filament assembly and stabilization. There is also extensive cross-talk between the Ras-ERK, PI3K-mTORC1, and PLC-γ1 pathways. The summary is based on data obtained

(*Continued*)

and for its medicinal properties for thousands of years. Marijuana could have been discussed in Chapter 16, Receptor-mediated mechanisms for drugs of abuse that may adversely affect brain development, which is devoted to the harmful effects of addictive substances on brain development. However, it was decided to describe marijuana's effects on the developing brain here because of its beneficial properties, including its interactions with BDNF.

Within the last 20 years there have been profound discoveries allowing for a great increase in our understanding of the mechanism of action of marijuana and its constituents. Studies have demonstrated the existence of endogenous cannabinoids (endocannabinoids, eCBs) in the body that primarily bind to two receptors. This discussion will concentrate on the roles of eCBs and their receptors in the brain, although they have major roles in other organs as well.

After the importance of eCBs and their receptors in the mature brain is described, the focus will turn to the roles of the eCBs in the developing brain. Finally the benefits and dangers of marijuana and its derivatives to both the developing and the mature brain will be summarized.

There are two major eCBs in brain: amantadine and 2-arachidonoyl glycerol. The synthetic and degradative pathways for each have been elucidated. These compounds are released after cleavage from both excitatory and inhibitory postsynaptic endings. After release, they bind to one of two G-protein coupled receptors (GPCRs) found in the brain, CB1R or CB2R, at the presynaptic endings. This binding triggers a pathway leading to the inhibition of further transmitter release (Fig. 10.8).

CB1R is much more abundant in the brain and is perhaps the most common GPCR found there. CB2R is found to a lesser extent in brain, but is much more common than CB1R in the rest of the body. Another GPCR, GPCR 55, has more recently been shown to bind eCBs and to be present in significant quantities in the brain as well. Its function, however, is not well understood. In addition to the type of eCB-mediated signaling described previously, there is eCB-mediated autocrine signaling at the postsynaptic ending as well, and also a postsynaptic neuron, astrocyte, presynaptic signaling.

◄ in different brain regions, experimental preparations, and paradigms. Many of the downstream pathways have not yet been shown to be regulated by endogenous BDNF in the context of L-LTP. Current evidence suggests that TrkB regulates multiple forms of mRNA-specific translation, controlled by CYFIP1, FMRP, and microRNAs, as well as more general regulation of translation initiation through mTORC1 and MNK. Dual regulation of translation and cytoskeletal dynamics by BDNF is another emerging theme. Reciprocal mechanisms may exist whereby cytoskeletal regulation promotes local translation, and translated proteins (such as Arc and LIMK1) influence actin filament assembly and stabilization. There is also extensive cross-talk between the Ras-ERK, PI3K-mTORC1, and PLCγ1 pathways.

Panja D, Bramham C. Neuropharm. 2014;76 (Pt C):664–76. Fig. 1. PMID: 23831365. With permission.

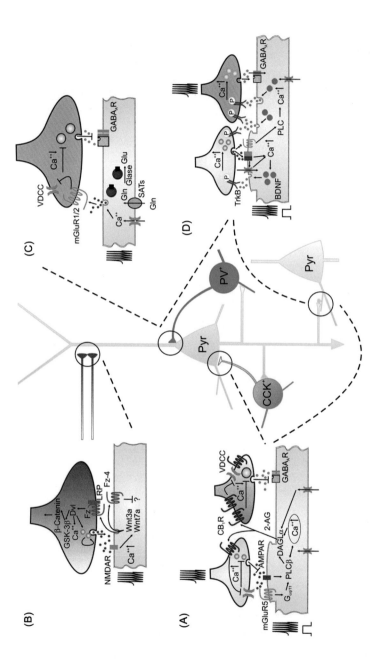

FIGURE 10.8 Overview of signaling mechanisms underscoring synaptogenesis in the embryonic CNS with continued control of retrograde neurotransmitter release at mature synapses.

(A) Activity-dependent eCB release inhibits neurotransmitter release from synapses of both pyramidal cells (Pyr) and cholecystokinin (CCK) + interneurons [2]. (B) Wnt signaling has been implicated in the control of presynaptic assembly and neurotransmitter release at excitatory afferents. (C) Dendritic release of glutamate provides negative feedback at perisomatic terminals of parvalbumin (PV) + basket cells. (D) By contrast, dendritic BDNF release enhances the efficacy of synaptic communication at select cortical synapses. *2-AG*, 2-arachidonoylglycerol; *AMPAR*, AMPA receptor; *CB1R*, CB1 cannabinoid receptor; *DAGLα/β*, sn-1-diacylglycerol lipase α/β isoforms; *Dvl*, disheveled; *Fz*, frizzled; *GABAAR*, GABAA receptor; *Gln*, glutamine; *Glase*, glutaminase; *GSK-3β*, glycogen synthase kinase-3β; *LRP*, low-density lipoprotein receptor; *mGluR*, metabotropic glutamate receptor; *NMDAR*, N-methyl-D-aspartate receptor; *PLC(β)*, phospholipase C (β isoform); *SATs*, system A amino acid transporters; *TrkB*, tyrosine kinase B receptor; *VDCC*, voltage-dependent Ca++ channel; *Wnt*, Wingless-Int family of ligands.

Harkany T, Mackie K, Doherty P. Wiring and firing neuronal networks: endocannabinoids take center stage. Curr Opin Neurobiol. 2008;18:338–45. Fig. 1. PMID: 18801434. With permission.

There is evidence that the eCBs and their receptors play key roles in determining the cell fate of neuronal precursors during very early neuronal specification. During the early decision of neuronal type discussed in Chapter 7, A testable theory to explain how the wiring of the brain occurs and Chapter 8, The development of the cerebral cortex, eCBs are critical (Fig. 10.9). Once their ultimate fates have been determined, many developing neurons express CB1R in their axonal and dendritic growth cones, which as described in Chapter 4, The growth cone: the key organelle in the steering of the neuronal processes; the synapse: role in neurotransmitter release, are dynamic neurite endings that sense and

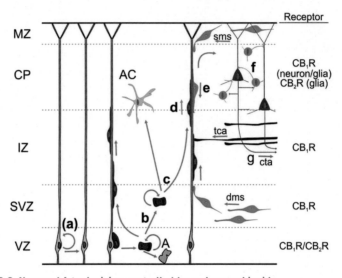

FIGURE 10.9 Neuronal fate decision controlled by endocannabinoids.

Studies with CB1/2R agonists and antagonists on cultured neurospheres and in adult mice have provided evidence that eCB signaling can affect neural stem cell fate and proliferation (A). During development, eCB signaling through CB1Rs affects neural progenitor proliferation (B) and lineage commitment (C) in the cortical subventricular zone (SVZ). Notably, CB1R expression is minimal on neuronal progenitors, whereas robust upregulation of CB1R expression coincides with neuronal commitment. Thus, eCBs have the potential to exert powerful control on both radially migrating postmitotic pyramidal cells (D; red) and tangentially migrating immature GABAergic interneurons (E; blue) populating the cortical plate (CP). Upon final positioning, eCB signaling contributes to the control of cell type—specific neuronal identification (F) and both intracortical and long-range axon patterning (G). *A*, apoptosis; *AC*, astrocyte; *CB1R*, type 1 cannabinoid receptor; *CB2R*, type 2 cannabinoid receptor; *cta*, corticothalamic axon; *dms/sms*, deep/superficial migratory stream; *IZ*, intermediate zone; *MZ*, marginal zone; *tca*, thalamocortical axon; *VZ*, ventricular zone.

Harkany T, Mackie K, Doherty P. Wiring and firing neuronal networks: endocannabinoids take center stage. Curr Opin Neurobiol. 2008;18:338—45. Fig. 2. PMID: 18801434. With permission.

respond to guidance cues. As shown in Fig. 10.9, growth cones predominantly contain *sn*-1-diacylglycerol lipase, which produces eCBs by cleavage from a phospholipid embedded in the plasma membrane. In contrast, the stabilized axon produces mainly monoglyceride lipase (MGL), which degrades eCBs. The lack of MGL expression in the growth cone allows it to produce eCBs, which in addition to eCB sources from neighboring cells, promote neurite outgrowth (Fig. 10.10). In contrast, in stabilized axons MGL degrades eCBs, blocking ectopic axonal CB1R signaling. Thus CB1R signaling occurs only in a defined cellular compartment, while the unwanted growth cone and off-target effects are prevented. MGL therefore serves a gatekeeper function in the growing axon, actively preventing the inappropriate activation of CB1Rs to allow a fine-spatial compartmentalization of eCB signaling (Fig. 10.10).

Finally, eCBs and their receptors can also provide target-derived molecules to participate in the final decisions leading to the formation of the appropriate synaptic connections between neurons. Therefore eCBs and their receptors can participate in all stages of neuronal development essential for the correct wiring of the brain.

With the great increase in the evidence demonstrating the importance of eCBs and their receptors in brain development during the last decade, a number of findings point to the interactions between neurotrophins and eCBs in this process. One example of the interactions between these two critical systems involves the migration of cholecystokinin containing interneurons from their site of origin in the ganglionic eminence to the hippocampus. It has been found that BDNF plays a key role in this process. Also eCBs act as tropic molecules for interneuron migration by binding to their own receptors and transactivating TrkB receptors.

BDNF and eCBs can also have yin−yang functions in nerve cells, both in early synaptogenesis and in mature neurons. BDNF, released by postsynaptic vesicle secretion, increases the probability of neurotransmitter release from the presynaptic cell. In contrast eCBs released from the postsynaptic terminal decrease the probability of neurotransmitter release from the presynaptic terminal.

The evidence described here indicates that eCBs and their receptors have important functions in both the developing and mature brain. Therefore components of marijuana that mimic eCBs can have profound effects on both. The non-hallucinogenic components are increasingly found to be useful in the treatment of a whole host of human ailments including anorexia, acute and chronic pain, sleeplessness, and nausea. Most importantly, components of marijuana that control pain are beginning to be used to replace opioids. This will in turn help alleviate the opioid epidemic the United States now faces.

As discussed previously, ECBs and their receptors play major roles in brain development. Therefore marijuana use by women, either during pregnancy or while attempting to become pregnant, may produce deleterious effects on the fetal brain. This is because the active components of marijuana are very lipophilic, can cross the placental barrier, and remain in the fetal brain for more than 6 weeks. It is also likely harmful for adolescents to use marijuana excessively since

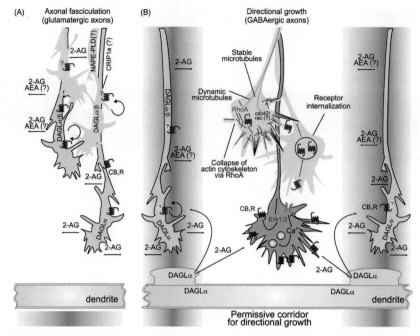

FIGURE 10.10 Unifying concept defines endocannabinoid-driven spatial segregation of inhibitory and excitatory synapses.

(A) Pyramidal cells emit the first axons in cerebral circuits. They exhibit moderate CB1R expression with these receptors distributed all along the axis of the elongating axon. *sn*-1-diacylglycerol lipases (DAGLα/β) are coexpressed in excitatory axons permitting cell-autonomous endocannabinoid (eCB) signaling that drives axonal elongation. Target-derived 2-arachidonoylglycerol (2-AG) might act as an additional attractive force. (B) Later-arriving GABAergic axons do not express DAGLs in their growth cones, but do express high levels of CB1Rs that sense a microenvironment that contains hotspots of 2-AG emanating from excitatory afferents (*red shading*). In these neurons, upon eCB stimulation, CB1Rs are removed from motile filopodia and translocated to the central growth cone domain where they activate the extracellular signal-regulated kinase 1/2 (Erk1/2) pathway, and RhoA GTPases. RhoA activation leads to a collapsing response on the side of the growth cone facing the eCB gradient thus contributing to growth cone steering decisions. Consequently, GABAergic axons target specific dendritic domains of postsynaptic neurons whose subcellular distribution and size are defined by excitatory afferents. *AEA*, 2-arachidonoylethaloamine/anandamide; *CB1R*, CB1 cannabinoid receptor; *CRIP1a*, cannabinoid receptor interacting protein 1a; *NAPE-PLD*, N-acyl-phosphatidylethanolamine-selective phospholipase D; *RhoA/Cdc42/Rac*, members of the Rho family of small GTPases.

Harkany T, Mackie K, Doherty P. Wiring and firing neuronal networks: endocannabinoids take center stage. Curr Opin Neurobiol. *2008;18:338–45. Fig. 3. PMID: 18801434. With permission.*

adolescent brains are still developing, and one well-documented side effect of marijuana use is short-term memory loss. Unfortunately, more scientific studies on both the beneficial and harmful effects of marijuana are needed given the legalization of recreational use in several states and of medical use in many others.

GENERAL REFERENCES

Recommended Book Chapters

Kandel ER, Schwartz JH, Jessel TM, Siegelbaum SA, Hudspeth AJ. *Principles of Neural Science*. 5th ed. New York: McGraw Hill; 2013. Chapters 53, 56, 66, 67.

Squire LR, Berg D, Bloom FB, Du Lac S, Ghosh A, Spitzer NC. *Fundamental Neuroscience*. 4th ed. Oxford, UK: Elsevier; 2013 [Chapters 18, 21, 44].

Review Articles

Cowan WM. Viktor Hamburger and Rita Levi-Montalcini: the path to the discovery of nerve growth factor. *Annu Rev Neurosci*. 2001;24:551–600.

Harkany T, Mackie K, Doherty P. Wiring and firing neuronal networks: endocannabinoids take center stage. *Curr Opin Neurobiol*. 2008;18:338–345.

Hasselmo ME, Stern CE. Current questions on space and time encoding. *Hippocampus*. 2015;25:744–752.

Park H, Poo MM. Neurotrophin regulation of neural circuit development and function. *Nat Rev Neurosci*. 2013;14:7–23.

Axon and dendrite guidance molecules and the extracellular matrix in brain development

11

CHAPTER OUTLINE

In this chapter, molecules that were first isolated based on their ability to act as axon guidance molecules are examined (Fig. 11.1). Several families of these molecules have this property but also play many other key roles in the developing brain. Finally, glycoproteins and proteoglycans, which are important in axon migration as well as in other developmental areas including the formation of the extracellular matrix (ECM) are discussed.

As described in Chapter 1, Classes of receptors, their signaling pathways, and their synthesis and transport, ligands can be divided into two types: secreted molecules and surface-bound molecules. In fact some families of molecules, including the Semaphorins, can be either a surface molecule or, if cleaved by a protease, become secreted molecules.

Another major division of ligands that affect the growth cone is determined by whether they function as repulsive or attractive. The Ephrins and the Semaphorins, whether surface bound or secreted, are usually repulsive. The members of the neurotrophin family are always attractive as long as they interact with their high-affinity receptors. The Netrin family members can be either repulsive or attractive, depending upon which type of receptor they bind to. There are many other tropic or repulsive molecules expressed in the brain but they will not be discussed here.

Since axons are on average much longer than dendrites, they must travel the correct pathway through a veritable jungle of neurons and their processes. Many families of axon guidance molecules and their receptors have been discovered since Levy-Montalcini first reported the isolation and characterization of nerve growth factor (NGF), for which she was awarded the Nobel Prize in Physiology and Medicine in 1986.

Just as NGF and other members of the neurotrophin family serve numerous functions besides axon guidance (see Chapter 10: The key roles of BDNF and endocannabinoids at various stages of brain development including neuronal commitment, migration, and synaptogenesis), the three families of receptors and

Receptors in the Evolution and Development of the Brain. DOI: https://doi.org/10.1016/B978-0-12-811012-6.00011-X

149

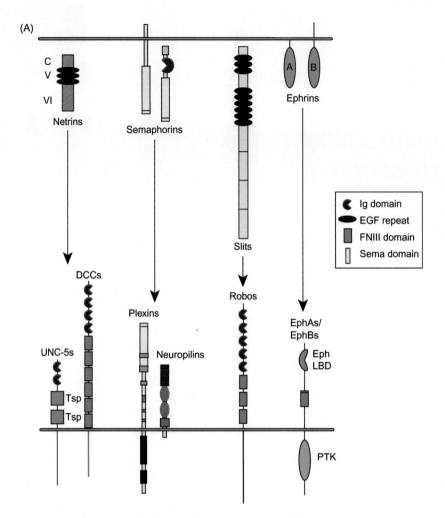

(A)

C
V
VI
Netrins

Semaphorins

Slits

A B
Ephrins

Ig domain
EGF repeat
FNIII domain
Sema domain

DCCs

Plexins

Robos

EphAs/
EphBs

UNC-5s

Neuropilins

Eph
LBD

Tsp
Tsp

PTK

(B)

	Cues	Receptors	Sign of response
Morphogens	Wnts	Frizzleds	+
		Derailed/Ryk	−
	HH	BOC + Smoothened	+
	Bmps	BMPR and BMPR-II	−
Growth factors	GDNF	GFRα and Ret	+
	FGFs	FGFRs	+
	HGF	Met	+
	Neurotrophins	Trks	+

FIGURE 11.1

Multiple families of molecules serve as guidance cues. (A) Structures of the "canonical" axon guidance cues: netrins, semaphorins, slits, ephrins, and their receptors. Netrin receptors include members of the DCC and UNC5 family. There are seven classes of semaphorins (two

(Continued)

ligands described herein also have many other key functions in both the developing and mature CNS. Fig. 11.1 shows examples of the three ligand—receptor systems discussed here: the Ephrin-EphRs, the Semaphorin-Plexins, and the Netrin-DCCs. Also morphogens, which can act as axon guidance molecules, are listed in this figure but are not discussed further.

Eph receptors (EphRs) are plasma membrane embedded tyrosine kinases that comprise the largest family of these important signaling receptors in the human genome. EphRs are divided into two classes: EphAs and EphBs (Fig. 11.1). EphRs dimerize and autophosphorylate after binding to their ligands, the Ephrins, as described in Chapter 1, Classes of receptors, their signaling pathways, and their synthesis and transport. The EphR autophosphorylation sets off a cascade of downstream events including rearranging the actin-containing cytoskeleton, which produces growth cone collapse leading to the growth cone changing directions. Ephrins are embedded in the plasma membrane, as are the EphRs; one class anchored via a transmembrane sequence, the other via a linkage to a glycosylphosphoinoside (GPI) group in the outer leaflet of the plasma membrane. Surprisingly the Ephrins can also signal, a process called reverse signaling. In the case of Ephrins having a transmembrane domain, the cytosolic sequence contains a PDZ domain that serves to cluster many different proteins. It is not completely understood how the GPI containing Ephrins signal. However, there is some evidence that they interact with and activate one or more tyrosine kinases that are anchored to the cytoplasmic leaflet of the plasma membrane via a phospholipid moiety (Fig. 11.1).

It is thought that to signal, a neuronal process expressing an EphR must bind to another neuron soma or process expressing an Ephrin. Since the EphR-Ephrin binding must occur on adjacent cells, it can only mediate short-range interactions. After this interaction occurs, an extracellular protease cleaves the Ephrin, thus allowing the Ephrin-containing cell to migrate in another direction.

The major benefit of the repulsive mechanism employed by the EphR-Ephrin system in axonal migration is to prevent axons from ceasing migration prematurely before reaching their appropriate targets. In contrast, other molecules such

---

◀ invertebrate and five vertebrate). Of the $\sim$20 mammalian semaphorins, 5 are secreted (class 3) and the others are divided into four families of transmembrane or GPI-linked proteins (represented here by a single diagram). Semaphorin receptors include plexins (of which nine are known in mammals, divided into classes A—D) and neuropilins (two in mammals). Class 3 semaphorin signaling is mediated by complexes of neuropilins (which serve as binding moieties) and plexins (signaling moieties). Slit receptors are members of the Roundabout (Robo) family. The receptors for Ephrins are members of the Eph family. Ephrin As are GPI-anchored and predominantly bind EphAs, whereas Ephrin Bs are transmembrane proteins and preferentially bind EphBs. (B) Morphogens (members of the Wnt, Hedgehog, and BMP families) and various growth factors listed here also serve to guide neuronal processes. Known receptors functioning to mediate their effects in guidance are indicated.

Squire LR, Berg D, Bloom FB, Du Lac S, Ghosh A, Spitzer NC. Fundamental Neuroscience. 4th ed. Oxford, UK: Elsevier; 2013 [Chapters 16, 54]. Fig. 16.5, p. 367. With permission.

as the Netrins discussed below can be attractive. They are important during axon guidance to bundle axons growing into the same target area.

The Ephrins and the EphRs were first characterized as molecules that play key roles in the guidance of retinal ganglion cell axons to their targets in the optic tectum. They also play key roles in the guidance of other axons in the CNS.

There are also many examples of members of the EphRs and Ephrins playing key roles in the development of the brain, beside their role in axon guidance. Here the discussion will be focused on their multiple functions during the development of the cerebral cortex, which was discussed in Chapter 8, The development of the cerebral cortex.

Many different Ephrins and EphRs are expressed at different stages of cortical development. During the earliest cell divisions in the ventricular zone (VZ), EphA4 and Ephrin B1 are expressed on adjacent cells and only these cells divide. Another receptor that is implicated in early VZ cell division, the fibroblast growth factor receptor, which can serve as a signaling coreceptor for EphRs, is also expressed in this unique population of cells. At the same stage of development, EphB2 is expressed in a more superficial layer of the developing cortex, the cortical plate. Its interaction with Ephrin B1−containing cells prevents them from differentiating, thus ensuring the continued cell division in the VZ.

During the subsequent stage of cortical development, at which time postdivision neurons destined to become excitatory neurons migrate away from the VZ along radial glial cells (RGCs), EphRs and Ephrins again play important roles. During this developmental stage, Ephrin B1 expression is blocked by a mechanism that involves microRNA binding to and inactivation of its mRNA. This inactivation allows neuronal differentiation.

After neuronal differentiation occurs, Ephrin A−mediated signaling directs radial migration of neuronal precursors along RGCs. Deleting Ephrin A3 through gene knockout causes premature detachment of these neurons from the RGCs.

As discussed previously, the first role described for the Ephrin-EphR system was in axon guidance. In the cerebral cortex during development, several guidance pathways are heavily influenced by this system. One striking example is the thalamocortical axons, which reach their target zone in layer 4 of the cortex just as the excitatory pyramidal cells finish differentiation. A gradient of Ephrin A5 in the subplate zone repels the EphA4 thalamic axons, and forces them to enter the cortex.

Another striking example of the key roles played by the EphR-Ephrin system in axon guidance occurs during GABAergic cell migration away from the ganglionic eminences. An Ephrin-A3 gradient in the ganglionic eminence repels newly born inhibitory neurons, routing them toward the cerebral cortex. Other Ephrin gradients direct corticothalamic axons toward their appropriate thalamic nuclei, while EphA8 and 5 spatially segregate the axons of the corpus callosum that cross from one half of the cortex to the other half.

The Ephrin-EphR system also plays a role in establishment of dendritic branches and spines that become postsynaptic endings. Overexpression of EphB2 in mice produces increased spine density, while gene knockout produces the opposite result.

Another vitally important role for the Ephrin-EphR system is in setting up the cortical compartments of the early developing brain. Different EphRs and Ephrins are sequentially expressed as these specialized compartments are formed. There is evidence that some Ephrin-EphR patterns are preserved in the absence of innervations while other patterns depend on correct innervations. Also Ephrin-EphR interactions are critical in setting up the topologic map of the RGC axonal synapses in the thalamus (Fig. 11.2).

Finally there is evidence that the Ephrin-EphRs are involved in determining the correct number of neurons in the brain. Overexpression of EphA7 stimulates apoptosis in the mouse brain leading to smaller brains in adults; whereas downregulation leads to adult mice having larger brains.

Semaphorins (Semas), which were first described as axon guidance molecules, constitute another large family. However, subsequent studies have revealed their key roles in many aspects of brain development. While Semas are found in many invertebrate species, this discussion will be confined to the development of the vertebrate brain. The vertebrate Sema family consists of 20 members including secretory molecules, molecules with a transmembrane domain, and molecules anchored to the plasma membrane by a GPI residue (Fig. 11.1).

Most Semas signal through their interaction with one of three members of the Plexin family but also concomitantly with a variety of coreceptors including Neuropilin (NRP), IgCAM, VEGFR2, Integrins, and two proteoglycans, heparan sulfate proteoglycan (HSPG) and chondroitin sulfate proteoglycan (CSPG). In the absence of NRP the Semas are repulsive for axon growth, while in its presence they become attractive. Also when Sema5A binds to a Plexin and a CSPG, the axon is repulsed while the same SemaA5-containing axon is attracted when a HSPG is contained in the multisubunit receptor (Fig. 11.3).

The SEMA-Plexin/NRP complex is also involved in neuronal migration. If the amounts of NRP1 and 2 are lowered by deletion of the trascription factor COUP-TF2, many GABAergic neurons are misrouted during their movement from the caudal ganglionic eminence to the amygdala.

Cis-interactions between SemaA6 and Plexins on the same neuron can render both molecules incapable of making *trans*-interactions. An example of this occurs in the developing hippocampus, when a *cis*-interaction between Sema6A and a Plexin on the same neuron blocks the pyramidal Sema6A, mossy fiber Plexin A4 interaction.

Semas are also involved in the establishment of axon and dendrite identity. It was originally believed that the axon is specified first and dendrites develop by a default pathway. However, there is now strong evidence that dendrite identity is also specified by a group of molecular signals including the Sema3A-NRP system. This process involves a gradient of Sema3A binding to NRP1, thus activating a cGMP kinase preferentially located in the cytosol underlying the apical dendrite. This in turn triggers a signaling sequence leading to the specification of the dendrite while axon identity is inhibited. At the other end of the differentiating neuron, where the level of Sema3A is low, the axon can be specified via increased cAMP.

While EphRs and Ephrins are critical to the topological mapping of the cerebral cortex during development, the Sema—Plexin interactions play critical

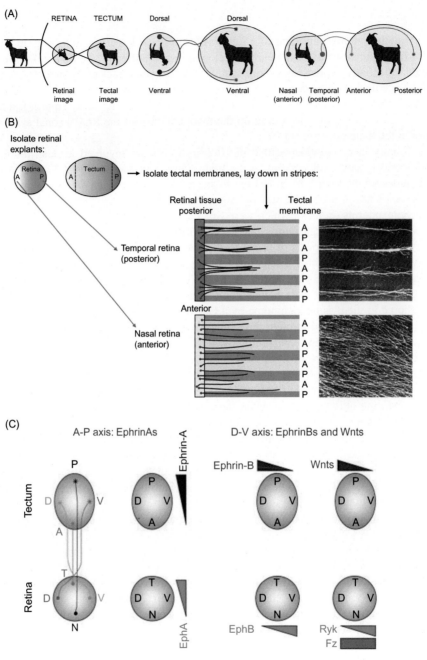

FIGURE 11.2

Topographic mapping of retinal projections in the visual system. (A) The projection of
retinal ganglion cells (RGCs) to the optic tectum is topographic. A map of the visual world
on the tectum or superior colliculus (left) is inverted with respect to that on the retina

(*Continued*)

roles in setting up the laminar organization of the neurites during development of the retina. The retina is composed of two layers or lamina of neurons: the inner nuclear layer and the ganglion cell layer. The synapses in the retina occur in two lamina. The inner plexiform layer (IPL) contains synapses between retinal ganglion cells, amacrine cells, and bipolar cells; while the outer plexiform layer (OPL) contains synapses between photoreceptor cells, horizontal cells, and bipolar cells. At present, between 50 and 100 different classes of neuron have been identified in the retina of vertebrates and more will certainly be identified. Each type of neuron identified so far resides in a particular sublamina and displays a specific connectional pattern.

During development of the retina, each neuron must send out neurites into either the IPL or the OPL. Then the neurites must target the appropriate sublaminar region and finally establish the proper synaptic connections. Recently Sema5A and B have been demonstrated to be present in the outer neuroblastic layer of retina during very early development. These proteins target their receptors (PLXNs 1 and 3) and repel them, causing their neurites to target only the IPL. Once these neurites are properly oriented, Sema6A is expressed in sublamina S4 and 5; while its receptor, PLXN A4 is expressed in sublamina 1−3. In normal mice, a group of amacrine cells containing tyrosine hydroxylase have neurites in S1. In contrast, mice deficient in either Sema6A or Plxn A4 have neurites extending into layers 4 and 5.

◀ because axons from dorsal and ventral retina project to ventral and dorsal tectum, respectively (middle), and axons from anterior (nasal) and posterior (temporal) retina project to posterior and anterior tectum, respectively (right). (B) An in vitro assay demonstrates topographically appropriate recognition of tectal membranes by some retinal axons. Small slivers of nasal and temporal retina (top left) are placed on carpets of membranes from tectal cells and allowed to grow. The carpets have alternating stripes of anterior and posterior tectal membranes. In these experiments, the posterior (temporal) retinal axons show an appropriate preference for growth on membranes derived from their correct target (anterior tectum). Other experiments showed that this is due to avoidance of a repellent activity, due to EphrinA ligands, that is enriched in posterior tectum. Note that in this experiment, anterior (nasal) axons do not show the expected preference for posterior tectal membranes, which represents a limitation of the assay; more refined assays are, however, able to demonstrate this. (C) A model for topographic mapping by gradients of guidance molecules and their receptors. Along the A-P axis, retinal neurons have a graded expression of EphA receptors, which enables them to interpret a decreasing posterior-to-anterior gradient of EphrinA ligands on the tectum to select their appropriate target on that axis. Mapping on the D-V axis, in contrast, involves gradients of EphrinB and Wnt ligands that are read by graded distributions of cognate receptors EphB and Frizzled, respectively. Wnt ligands are also detected by Ryk, which has a nongraded distribution.

Squire LR, Berg D, Bloom FB, Du Lac S, Ghosh A, Spitzer NC. Fundamental Neuroscience. *4th ed. Oxford, UK: Elsevier; 2013 [Chapters 16, 54]. Fig. 16.13, p. 380. With permission.*

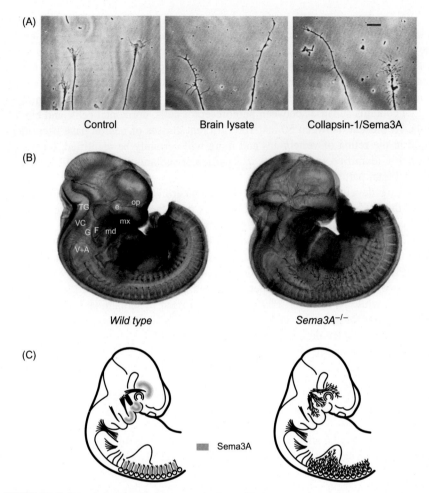

FIGURE 11.3 Secreted semaphorins are potent repellents.

(A) Dorsal root ganglion neurons in culture exhibit a dramatic collapse when exposed to biochemically enriched fractions derived from chick brain lysates or to conditioned medium from tissue culture cells engineered to express recombinant Sema3A/collapsin-1. Medium from control nontransfected cells has no effect on DRG growth cone morphology. (B) Mouse whole-mount, embryonic day 11.5, wild-type and Sema3A mutant embryos immunostained for neurofilaments to illuminate spinal and cranial peripheral nerve projections. In Sema3A mutants, cranial nerves, including the trigeminal nerve (TG), are dramatically defasciculated, extend branches into ectopic locations, and overshoot their normal target regions. Spinal nerves, too, are defasciculated and extend into intersomitic regions they normally avoid (cranial nerves: *op*, ophthalmic; *mx*, maxillary; *md*, mandibular; *F*, facial; *G*, glossopharyngeal; *VC*, vestibulocochlear; *V + A*, vagus and accessory; *e*, eye. Spinal nerves: *arrow*). (C) Schematic of whole-mount embryos in (B) showing cranial and spinal nerve projections, and also Sema3A localization (*gray*).

Squire LR, Berg D, Bloom FB, Du Lac S, Ghosh A, Spitzer NC. Fundamental Neuroscience. *4th ed. Oxford, UK: Elsevier; 2013 [Chapters 16, 54]. Fig. 16.8, p. 371. With permission.*

Finally Semas, Plexins, and NRPs play important roles in synaptic specificity in the developing cerebral cortex. In normal mice, the apical dendrites of pyramidal cells contain very few dendritic spines and postsynaptic specializations. However, mice with the gene for either Sema 3F, NRP2, or PLXN3 deleted contain more spines on the apical dendrites that form during development.

The Netrins were also first discovered as axon guidance molecules, but then were determined to be involved in many other processes occurring during brain development. The Netrins, five of which are found in mammals, are members of the Laminin family of ECM molecules (see discussion of Laminin in Chapter 1: Classes of receptors, their signaling pathways, and their synthesis and transport). Netrins 1, 3, and 4 are secreted and can become part of the ECM, while Netrins G1 and G2 are attached to the plasma membrane by a GPI residue.

Four classes of Netrin receptors have been identified in mammals. These include two members of the DCC family, Neurogenin and DCC; four members of the UNC family, UNC5A−D; DESCAM; and NGL1 and 2. All of these receptors are one-pass transmembrane plasma membrane−bound receptors that are members of the immunoglobulin superfamily (see Fig. 11.1). The first three receptor classes bind to the secreted Netrins while the NGLs bind only to the GPI linked netrins.

Secreted Netrins can serve as tropic molecules for several types of embryonic neurons including retinal ganglion cells and motor neurons through their binding to the DCC receptor or its closely related homologue, Neurogenin. In contrast, members of the Unc5 family, when bound to a secreted neurogenin in frog spinal neurons, appear to be repellent. An example of this action is found with UNC5 homologues, which are repelled by Netrins. In some cases the repellent action requires that the neuron expresses both an UNC5 and a DCC (Fig. 11.4).

Netrins are also involved in early neuronal precursor cell migration. Precerebellar neuronal cell migration is directed by Netrin 1 secretion. However, during postnatal development of the cerebellum, Netrin 1 repels cerebellar granule cell precursors that express UNC5. Netrins also play an important role in the maturation of oligodendrocytes, the formation of myelin, and the maintenance of the nodes of Ranvier.

Finally, Netrins play key roles in axon branching and synaptogenesis. Netrin secretion stimulates the polymerization of actin at sites that become branch points. Also perfusion of Netrin 1 into the frog optic tectum during embryogenesis produces more axon branching and more presynaptic specializations.

Finally, a fundamental process in neurite guidance is the avoidance of a neuron's own dendrites. A key family of proteins that participate in this self/nonself determination are the protocadherins (PCDHs). While these proteins will be discussed in detail in Chapter 12, Key receptors involved in laminar and terminal specification and synapse construction, with respect to their roles in the final neuronal specification, they also play an important role in axonal guidance. Neurons missing their correct complement of PCDHs have tangled dendrites that cross rather than avoid one another.

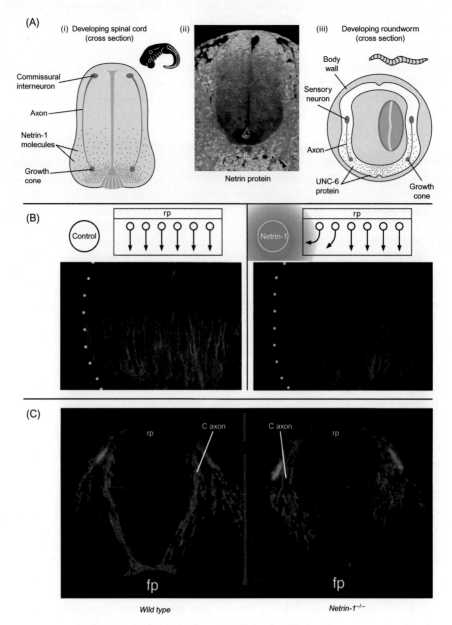

FIGURE 11.4 Netrins are phylogenetically conserved axon attractants.

(A) Distribution and function of Netrins in axon guidance. (i) Diagram of a cross-section through the embryonic vertebrate spinal cord depicting the role of Netrin-1 (here and in all subsequent schematic figures shown in *green*) in guiding commissural axons from the dorsal spinal cord to the ventral midline. (ii) A ventral to dorsal gradient of Netrin protein in the developing mouse spinal cord visualized using an immunohistochemical stain. (iii) Diagram of a cross-section through the developing nematode *Caenorhabditis elegans* illustrating the role of the Netrin/UNC-6 in guiding axons along a dorsoventral trajectory. (B) Attractive effect of Netrin-1 on rodent commissural axons in an in vitro assay for

(Continued)

Besides the four examples provided above there are numerous other neurite guidance proteins that have been or will be discovered. It is likely that all of them will be shown to have other roles in neuronal development as well as in the mature brain. Also, mutations in these proteins are very likely to have negative consequences for brain development and/or function.

Glycoproteins and proteoglycans also play major roles in brain development. Glycoproteins and proteoglycans both are protein- and complex carbohydrate–containing molecules. Glycoproteins contain most of their mass as protein while proteoglycans contain mostly carbohydrate. As discussed in Chapter 1, Classes of receptors, their signaling pathways, and their synthesis and transport, the oligosaccharides on glycoproteins are attached to Asn residues, while the oligosaccharides on proteoglycans are attached to Ser or Thr residues.

Glycoproteins and proteoglycans constitute a large percentage of the material that forms the ECM. In all organs including the brain, both constituents play key roles in retaining water and ions by cross-linking with other ECM components. Collagen, fibronectin, and laminin are major components in this three-dimensional matrix, both during and after development. In the case of the brain, this matrix surrounds both neurons and glia. The brain, unlike other organs, contains very few collagen fibers except those surrounding blood vessels. Instead glycoproteins and proteoglycans, together with the fiber forming proteins laminin and fibronectin, constitute the major components of the ECM.

Proteoglycans can be divided into two major categories: membrane associated and secreted. Most of the carbohydrate-containing portions of both, which are attached to Ser or Thr residues, contain many sulfated carbohydrates (Fig. 11.5).

◀ chemoattraction. Pieces of the dorsal half of the spinal cord, viewed from the side, were placed in culture for 40 h with either control COS cells or COS cells secreting Netrin-1. In control conditions (left), all the commissural axons grow from their cell bodies at the top (dorsal) side of the explants along a vertical (dorsoventral) trajectory [shown in diagrammatic form at the top, and visualized within the spinal cord explants using an immunohistochemical axonal marker (*red*) at bottom]. In the presence of cells secreting Netrin-1, however, the axons within ∼150 μm of the cells (*arrowheads*) turn toward the Netrin source. Dots mark the boundary between COS cells and spinal cord tissue. In the presence of Netrin-1 axons can be seen invading the Netrin-secreting COS cell clump. (C) Netrin-1 is required for guidance of commissural axons to the ventral midline. Shown are cross-sections through the spinal cord of embryonic day 11.5 mice. Left: wild-type (control); right: Netrin-1 knock-out mouse. Axons are visualized with an axonal marker. Normally (left), commissural axons (c) extend along a smooth dorsal-to ventral trajectory to the floor plate (fp) at the ventral midline. In the absence of Netrin-1 (right), they extend normally in the dorsal half of the spinal cord, but when their trajectory through the ventral half is profoundly perturbed, with many projecting medially or laterally, only a very few axons reach the floor plate.

Squire LR, Berg D, Bloom FB, Du Lac S, Ghosh A, Spitzer NC. Fundamental Neuroscience. 4th ed. Oxford, UK: Elsevier; 2013 [Chapters 16, 54]. Fig. 16.6, p. 368–369. With permission.

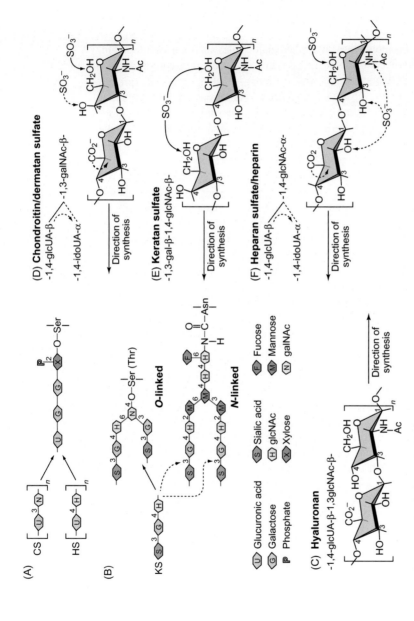

FIGURE 11.5 Synthesis of glycosaminoglycans (GAGS).

(A and B) Three short oligosaccharides link GAGs (left) to proteoglycan core proteins (right). (A) A tetrasaccharide anchors chondroitin sulfate (CS), dermatan sulfate, and heparan sulfate (HS) to serine residues. (B) Two different, branched oligosaccharides link keratan sulfate (KS) to either serine or asparagine. (C–F) Four parent polymers and postsynthetic modifications. (C) Hyaluronan [D-glucuronic acid β (1→3) D-N-acetylglucosamine β (1→4)]n (n ≥ 25,000) is not modified postsynthetically. (D) Chondroitin sulfate and dermatan sulfate are synthesized as [D-glucuronic acid β (1→3) D-N-acetylgalactosamine β (1→4)]n (n usually <250) and then modified. Some N-acetylgalactosamines are sulfated. In dermatan sulfate, D-glucuronic acids are epimerized to L-iduronic acid. (E) Keratan sulfate is synthesized as [D-galactose β (1→4) D-N-acetylglucosamine β (1→3)]n (n usually = 20–40) and then modified by sulfation. (F) Heparan sulfate/heparin is synthesized as [D-glucuronic acid β (1→4) D-N-acetylglucosamine α (1→4)]n (n usually <100) and then modified by sulfation and by epimerization of D-glucuronic acid to L-iduronic acid. galNAc, N-acetylgalactosamine; glcNAc, N-acetylglucosamine.

Pollard TD, Earnshaw WC, Lippincott-Schwartz J, Johnson GT. Cell Biology. 3rd ed. Philadelphia, PA: Elsevier; 2016. Fig. 46.18, p. 811. With permission.

Proteoglycans bind other ECM molecules, trophic factors, etc., through both their protein cores as well as through their glycosaminoglycans.

Surface-bound proteoglycans serve as direct receptors for a number of growth factors that serve as migration factors during brain development. Secreted proteoglycans form large aggregates in which growth factors as well as surface-acting proteases including ECM proteases are sequestered and protected from degradation.

As discussed in Chapter 8, The development of the cerebral cortex, proteoglycans play key roles in the radial migration of excitatory neurons in the developing cerebral cortex. A specific proteoglycan, pleiotrophin, is deposited along radial glial fibers and is important in pyramidal cell migration to the appropriate laminar layer. Also it is important in the polarization of each neuron from a multipolar entity to one having only one axon and several dendrites. Defects in pleiotrophin and other proteoglycans are responsible for a number of genetic diseases characterized by impaired neural migration.

Proteoglycans also play key roles in the tangential migration of inhibitory GABAergic interneurons from the median eminences to the cerebral cortex. A variety of secreted proteoglycans are expressed along the migratory route followed by these cells, in the striatum, as well as in the marginal zone, subplate, and subventricular zone of the developing cerebral cortex. Together with Sema 3A they provide important attractive and repulsive cues for these cells during their long journey.

In contrast to the large amount of evidence regarding the roles of various proteoglycans in brain development, there is at present much less information regarding contributions of glycoproteins. A major reason is that although more than 50% of the proteins made in the brain are glycosylated on Asn residues, they are almost all low-abundance components and therefore difficult to study. One glycoprotein that has been studied in great detail regarding its roles in brain development as well as in other processes is Reelin. This glycoprotein was discussed in detail in Chapter 8, The development of the cerebral cortex.

RECOMMENDED BOOK CHAPTERS

Kandel ER, Schwartz JH, Jessel TM, Siegelbaum SA, Hudspeth AJ. *Principles of Neural Science*. 5th ed. New York: McGraw Hill; 2013 [Chapter 54].

Pollard TD, Earnshaw WC, Lippincott-Schwartz J, Johnson GT. *Cell Biology*. 3rd ed. Philadelphia, PA: Elsevier; 2017 [Chapter 29].

Squire LR, Berg D, Bloom FB, Du Lac S, Ghosh A, Spitzer NC. *Fundamental Neuroscience*. 4th ed. Oxford, UK: Elsevier; 2013 [Chapters 16, 54].

Review Articles

Maeda N. Proteoglycans and neuronal migration in the cerebral cortex during development and disease. *Front Neurosci*. 2015;23(9):98.

Manitt C, Nikolakopoulou AM, Almario DR, Nguyen SA, Cohen-Cory S. Netrin participates in the development of retinotectal synaptic connectivity by modulating axon arborization and synapse formation in the developing brain. *J Neurosci*. 2009;29:11065−11077.

Pasterkamp RJ. Getting neural circuits into shape with semaphorins. *Nat Rev Neurosci*. 2012;13:605−618.

Key receptors involved in laminar and terminal specification and synapse construction

CHAPTER OUTLINE

The final stage of neuronal development in different regions of the brain involves the precise connection of each neuron with the appropriate neuron(s) to form a functional wiring network. As these connections assemble throughout the brain, the correct circuitry emerges.

After the neuronal growth cones near their targets, they must first find the appropriate area such as a lamina in the cerebral cortex or in the retina. There is good evidence that in the mammalian CNS, classical cadherins play key roles in this process. As shown in Fig. 12.1, the production of different cadherin isoforms that form homophilic synaptic adhesions with adjacent neuronal cells helps determine the terminal wiring diagram of the brain.

The protocadherin family of surface receptors are strong candidate for the molecules that produce these precise synaptic connections. This family was discussed briefly in the previous chapter in regard to its role in preventing the dendrites of an individual neuron from getting tangled up with other dendrites of the same cell.

There are over 80 protocadherin (Pch) genes in the human genome, including 12 Pch-α, 22 Pch-β, and 19 Pch-γ genes, which form a gene cluster. The cytoplasmic domain of each protein is identical and can bind to two important signaling tyrosine kinases: FAK and PYK2. The extracellular portion of each protocadherin contains five similar but nonidentical cadherin-like domains. These domains allow each molecule to bind to an identical molecule. Analysis of neurons at the single-cell level demonstrates that each neuron expresses 2 of the 12 Pch-α proteins, 4 of the 22 Pch-β proteins, and 4 of the 19 Pch-γ proteins in a randomly determined manner (Fig. 12.2).

Within each neuron, the expressed isoforms combine to form a tetramer and these tetramers in turn can only combine with the identical tetramer expressed on the synaptic membrane of another neuron. Therefore 78 combinations produced from the 12 Pch-α proteins, 26,796 from the 22 Pch-β isoforms, and 14,706 from

Receptors in the Evolution and Development of the Brain. DOI: https://doi.org/10.1016/B978-0-12-811012-6.00012-1

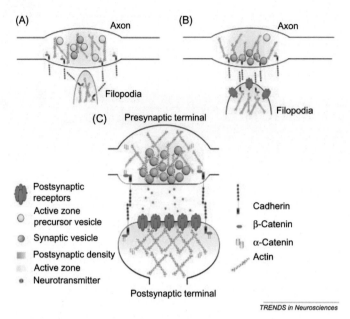

FIGURE 12.1 Schematic representation of hippocampal excitatory synapse formation and localization of the cadherin–catenin complex at different stages.

(A) Synaptogenesis is initiated by the formation of nascent contacts between dendritic filopodia and axons. At these stages, N-cadherin in distributed evenly along the synaptic structure. This is followed by (B) contact stabilization and clustering of cadherin and (C) maturity. A mature synapse is characterized by a stable presynaptic terminal that contains vesicles primed for neurotransmitter release and a postsynaptic density that contains the receptors and scaffolding proteins. In the adult, cadherin is localized to distinct regions bordering the mature active zone, termed as the puncta adherentia. Both β-catenin and α-catenin are known to colocalize with cadherin at these junctions. Recent studies in nonneuronal cells indicate that the binding of α-catenin to β-catenin and actin are mutually exclusive, with the monomer form having a higher affinity for β-catenin and the dimer having a higher affinity for actin. Although p120ctn and δ-catenin are known to be at synapses, their localization at the electron microscope level and during development is unclear.

Arikkath J, Reichardt LF. Cadherins and catenins at synapses: roles in synaptogenesis and synaptic plasticity.
Trends Neurosci. 2008;31:487–494. PMID: 18684518. Fig. 1. With permission.

the 19 Pch-γ isoforms can be generated from an individual neuron. Multiplying $78 \times 26{,}796 \times 14{,}706$ equals approximately 3×10^{10} combinations in the neurons of the brain. These combination are enough to account for the necessary variability to generate the many millions of unique neuronal categories found in the human brain.

Besides contributing to neuronal diversity during final synapse formation, clustered Pchs play central roles in other processes that are important to brain

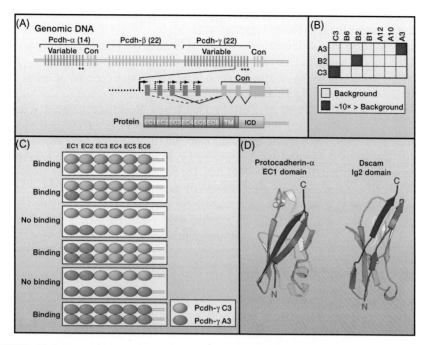

FIGURE 12.2 The protocadherin gene cluster and its protein products.

(A) The Pcdh gene cluster contains exons that encode 58 extracellular and transmembrane domains—14 in the α group (*purple*) and 22 each in the β (*orange*) and γ (*blue*) groups. Each ectodomain contains 6 cadherin repeats. Ectodomains are more related to others within a group than to those in other groups with the exception of αC1, αC2, and γC3–5 domains (*asterisks*), which are more closely related to each other than to neighboring members within their group. Each ectodomain is preceded by a promoter. Alternative splicing joins an α or γ ectodomain/transmembrane exon to three constant exons in the group. β exons encode complete proteins with short intracellular domains. (B) Results of adhesion assays in which each of seven Pcdh-γs was tested for binding to three isoforms. Each isoform bound preferentially to itself. (C) Cadherin domains EC2 and EC3 mediate the specificity of homophilic binding between isoforms. (D) Crystal structures of the EC1 domain of Pcdh-α and immunoglobulin domain 7 of Dscam showing the overall similarity of the β sandwich structure of Ig and cadherin repeats.

Zipursky SL, Sanes JR. Chemoaffinity revisited: dscams, protocadherins, and neural circuit assembly. Cell. 2010;143:343–353. PMID: 21029858. Fig. 3. With permission.

development. One postulated function of the Pchs is to participate in the production of the appropriate number of neurons in various regions of the brain. This results from the fact that if a neuron expressing a particular tetramer of Pchs cannot find a partner with the identical Pch tetramer with which to synapse, it will die due to lack of the necessary trophic support as discussed in Chapter 9, The importance of rapid eye movement sleep and other forms of sleep in selecting the appropriate neuronal circuitry; programmed cell death/apoptosis.

Another class of molecule that contributes to the generation of many unique neuronal types is the odorant receptors (ORs) found in the nasal passages. The ORs are capable of discriminating at least 100,000 distinct odors. The ORs constitute the most abundant class of G-protein coupled receptors in the genome, with at least 1000 found in the human genome. Any given odor consists of molecules that bind to a number of different ORs.

OR genes were isolated in 1991 by Drs. Linda Buck and Richard Axel, a discovery that led to them to be awarded the Nobel Prize in Physiology and Medicine in 2004. Each nasal epithelial cell (NEC), a specialized neuronal cell that is constantly dying and being replaced by an identical cell, contains only a number of identical ORs.

Each NEC responds with great sensitivity and selectivity to a unique odor. Also the identical group of ORs is found on the surface of the NEC axon, which grows into the olfactory bulb and synapses with a specific neuron (Fig. 12.3). Either deletion or overexpression of the gene for a particular OR in a NEC

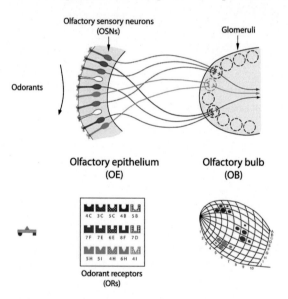

FIGURE 12.3 Conversion of olfactory signals.

In the OE, each OSN expresses only one functional OR gene in a monoallelic manner, which is referred to as the one neuron—one receptor rule. Furthermore, OSN axons expressing the same OR species converge to a specific glomerulus in the OB. Thus, each glomerulus represents one OR species. Since a given odorant activates multiple OR species and a given OR responds to multiple odorants, odorant signals received by OSNs in the OE are converted to a two-dimensional map of activated glomeruli with varying magnitudes of activity in the OB.

Sakano H. Neural map formation in the mouse olfactory system. Neuron. *2010;67:530–542. PMID: 20797531. Fig. 1. With permission.*

disrupts the connection between the NEC and the neuron in the olfactory bulb upon which it normally synapses.

The olfactory bulb consists of a series of glomeruli, that is, a closely grouped set of neurons. Each glomerulus receives synaptic input from a number of NECs that express the same OR (Fig. 12.3). The ORs on a given NEC play a key role in directing its axon to the correct glomerular location. However, it is not a direct role. It appears that each NEC OR produces a given level of cAMP. This in turn stimulates a group of attractive and repulsive molecules that connect axons containing identical ORs to form a bundle. Later cAMP increases the synthesis of another group of attractive and repulsive molecules in an activity-dependent manner. Some of the molecules that have been identified are members of the Sema, Slit, Robo, and Eph families and their respective receptors (Fig. 12.4). There is also evidence that members of the Pch and Cadherin families are involved in making the appropriate synaptic connections with their olfactory bulb target cells and maintaining the arrangement of the olfactory bulb cells respectively.

While some NEC olfactory bulb circuits are hardwired, others can be modified by experience during development and to a lesser extent throughout life. For example, there appear to be two types of fear responses to a predator's odor. One is apparently genetically determined and goes from the olfactory bulb directly to

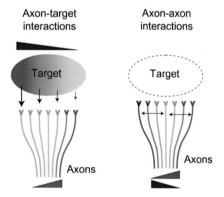

FIGURE 12.4 Preservation of the relative spatial relationships of projecting axons between origin and target sites is a general feature of neural map formation.

In the vertebrate visual system, graded expression of Eph receptors on retinal axons and their repulsive ligands, ephrins, from the superior colliculus regulates axonal projection of RGCs (left). Does map formation solely depend on axon–target interaction? In the mouse olfactory system, topographic order emerges in axon bundles well before they reach the target, suggesting pretarget axon–axon interactions. An axon guidance receptor, Nrp1, and its repulsive ligand, Sema3A, are both expressed by OSNs in the OE in a graded and complementary manner and mediate pretarget axon sorting within the bundles (right).

Sakano H. Neural map formation in the mouse olfactory system. Neuron. 2010;67:530–542. PMID: 20797531. Fig. 2. With permission.

the cerebral cortex. The other, which is learned, involves a memory component provided by the hippocampus (Fig. 12.5).

Once the ORs, Pchs, and other surface receptors make connections with their targets, both the pre- and postsynaptic terminal specializations must be constructed. Two key surface proteins that play major roles in this assembly are the presynaptic neurexins, which specifically bind to the postsynaptic neuroligins. These proteins were previously described in Chapter 4, The growth cone: the key organelle in the steering of the neuronal processes; the synapse: role in neurotransmitter release.

The neurexins consist of three family members that are expressed only on presynaptic terminals. Neurexins were first discovered as receptors for a neurotoxin, latrotoxin, which produces a massive neurotransmitter release. There are a number of different splice sites in the mRNA precursors for each of the three neurexin genes, which leads to the possibility of thousands of mRNAs and protein molecules. Deletion of all three family members is lethal. Neurexins contain several EGF domains on the extracellular portion of the molecule, a single transmembrane domain and a PDZ binding domain on the small cytosolic portion. Expression of neurexin on a nonneuronal cell induces a synapse-like structure. Also expression on a neuron produces a significant increase in synapse number. Once an axonal neurexin binds to its postsynaptic partner, usually neuroligin, surface receptors, Ca^{++} and Na^{+} channels, and other essential components of the presynaptic ending are assembled (Fig. 12.6).

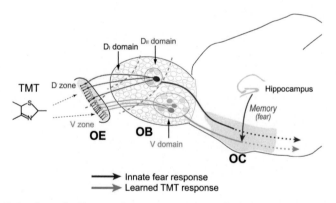

FIGURE 12.5 A schematic diagram of the neural pathways in the mouse brain.

A predator's odorant, TMT, activates two sets of glomeruli, one in the D domain and the other in the V domain of the OB. It has been proposed that TMT activates two different neuronal pathways: one for the innate fear response (*red*) and the other for the learned fear response based on memory (*green*). *OB*, olfactory bulb; *OC*, olfactory cortex; *OE*, olfactory epithelium.

Sakano H. Neural map formation in the mouse olfactory system. Neuron. *2010;67:530–542. PMID: 20797531. Fig. 7. With permission.*

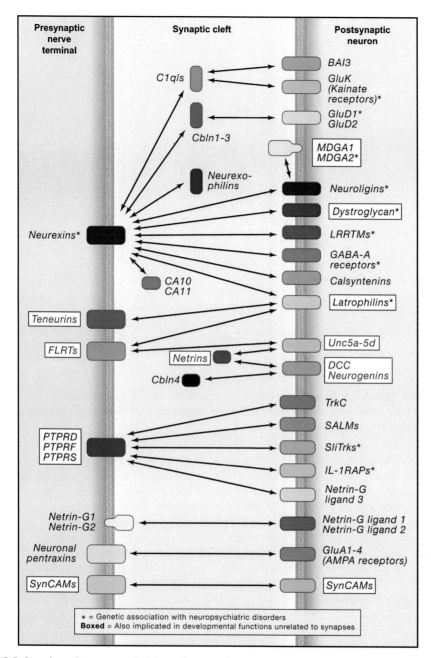

FIGURE 12.6 Overview of transsynaptic interaction complexes.

Proteins are shown schematically; arrows signify physical binding. Interactions that are specific to particular isoforms, such as neurexin splice forms or LAR-type PTPR variants, are not shown, and interactions shown may not apply to all members of a protein family. Although only selected interactions are shown that were chosen based on the level of evidence, significant uncertainty still exists about the validity of some of the interactions shown, and even the pre- versus postsynaptic localizations of some of the proteins (shown in *red typeface*) have not be definitively established.

Südhof TC. Synaptic neurexin complexes: a molecular code for the logic of neural circuits. Cell. 2017;171:745–769. PMID: 29100073. Fig. 12.2. With permission.

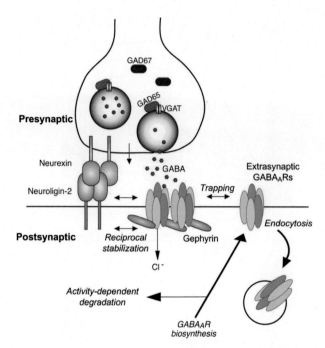

FIGURE 12.7 A hypothetical model depicting how GABA—GABA$_A$R signaling and neuroligin—neurexin adhesion may interact and cooperate to regulate the development of inhibitory synapses.

Pentameric GABA$_A$Rs are assembled in the endoplasmic reticulum and are subject to activity-dependent proteasomal degradation. Most GABA$_A$Rs are first delivered to extrasynaptic locations; they then either diffuse to and become trapped at postsynaptic sites or undergo endocytosis. NL2 and synaptic GABA$_A$R stabilize each other, either through intracellular reciprocal interactions aided by scaffolding proteins such as gephyrin or through extracellular *cis* interaction. In addition, GABA activation of GABA$_A$Rs might further stabilize synaptic GABA$_A$Rs through structural changes or signaling mechanisms. Such activity and GABA-mediated stabilization of GABA$_A$R might further increase the levels of NL2 at cell—cell contacts and, in turn, stabilize presynaptic terminals through transsynaptic interactions with neurexins.

Huang ZJ, Scheiffele P. *GABA and neuroligin signaling: linking synaptic activity and adhesion in inhibitory synapse development.* Curr Opin Neurobiol. *2008;18:77—83. PMID: 18513949. With permission.*

The neuroligin (NLGN) family, which consists of four similar members, possesses an extracellular acetylcholinesterase-like domain, a transmembrane sequence, and a short cytoplasmic PDZ binding domain. NLGN1 is localized at excitatory postsynaptic endings, NGLN2 at inhibitory postsynaptic endings, NLGN3 at both types of postsynaptic endings; and the most distant family member, NLGN4 at inhibitory postsynaptic endings. The NLGNs function as dimers, or in some cases, multimers.

The NLGN1 cytoplasmic PDZ binding domain binds to PSD-95, the first PDZ domain containing molecule to be discovered. PSD-95, a key scaffolding protein, is present in glutaminergic postsynaptic endings. NLGN1 also binds to other PDZ containing scaffolding proteins during the assembly of the postsynaptic endings.

NLGN2 binds to and stabilizes the cytosolic scaffold-forming protein, gephryn, at early inhibitory postsynaptic specializations. NLGN2 also binds to the membrane protein collybistin, which in turn recruits gephryn to the plasma membrane. In turn, gephryn forms multimers and clusters GABA receptors (Fig. 12.7).

Another rather unexpected protein takes part in initial synapse formation. It has recently been shown that the TrkC receptor, whose ligand is NT3, one of the neurotrophins, enhances excitatory synapse formation through its interaction with the protein tyrosine phosphatase sigma (PTPσ). This interaction occurs at a separate site on TrkC than that for NT3 (Fig. 12.6). Interestingly, NT3 binding to TrkC enhances TrkC-PTPσ mediated synapse formation. Perhaps other trophic factor receptors will be discovered to play this additional role during synaptogenesis.

GENERAL REFERENCES

Recommended Book Chapters

Kandel ER, Schwartz JH, Jessel TM, Siegelbaum SA, Hudspeth AJ. *Principles of Neural Science*. 5th ed. New York: McGraw Hill; 2013 [Chapters 32, 55].

Squire LR, Berg D, Bloom FB, Du Lac S, Ghosh A, Spitzer NC. *Fundamental Neuroscience*. 4th ed. Oxford, UK: Elsevier; 2013 [Chapter 17, 23].

Review Articles

Arikkath J, Reichardt LF. Cadherins and catenins at synapses: roles in synaptogenesis and synaptic plasticity. *Trends Neurosci*. 2008;31:487–494.

Huang ZJ, Scheiffele P. GABA and neuroligin signaling: linking synaptic activity and adhesion in inhibitory synapse development. *Curr Opin Neurobiol*. 2008;18:77–83.

Sakano H. Neural map formation in the mouse olfactory system. *Neuron*. 2010;67:530–542.

Südhof TC. Synaptic neurexin complexes: a molecular code for the logic of neural circuits. *Cell*. 2017;171:745–769.

Zipursky SL, Sanes JR. Chemoaffinity revisited: dscams, protocadherins, and neural circuit assembly. *Cell*. 2010;143:343–353.

Steroid hormones, glucocorticoids, and the hypothalamus

13

CHAPTER OUTLINE

13.1 STEROID HORMONES

In a given mammalian species, having two X chromosomes determines that the individual will be a female; while possessing one X and one Y chromosome causes the individual to be male. At birth the male and female brains are very similar; but shortly after birth sex hormones produced by either the male testicles or the female ovaries begin to circulate through the bloodstream and enter the cells of the brain. The ability for these molecules to cross the blood–brain barrier is due to the steroid sex hormones being synthesized from the extremely hydrophobic molecule cholesterol. The testes produce mostly testosterone, while the ovaries produce mostly estrogen, which is synthesized from testosterone.

Interestingly, testosterone is also the precursor to 5-alpha-hydroxy testosterone (DHT), which binds about three times more tightly to the testosterone receptor than does testosterone itself. If the targeted neurons contain the enzyme aromatase, estrogen is produced. In contrast, if the targeted neurons contain the enzyme, 5-alpha-reductase, then DHT is produced.

Together testosterone and DHT are referred to as androgens and their receptor as the androgen receptor. These receptors are produced in targeted regions of the developing brain in males. These areas include regions of the hypothalamus, striatum, and amygdala. Once testosterone enters neurons containing these proteins, some of the testosterone is converted to the more potent DHT. Both proteins bind to the testosterone receptor in the nucleus. The activated receptors form dimers and turn on genes associated with male behaviors.

In other regions of the brain, including the septum, hippocampus, hypothalamus, and midbrain, nuclear estrogen receptors are present. After estrogen binding, these receptors also dimerize, turning on genes associated with female behaviors.

Receptors in the Evolution and Development of the Brain. DOI: https://doi.org/10.1016/B978-0-12-811012-6.00013-3

The following examples illustrate how the steroid sex hormones function as trophic factors for particular groups of central nervous system (CNS) neurons during development. Binding to their specific receptors in targeted regions results in behaviors that are characteristically male or female.

A CNS motor circuit is present in the spinal nucleus of the bulbocavernosus muscle of most mammals. This neuromuscular system controls the penile erection and various vaginal movements. During early development, the number of motor neurons in the bulbocavernosus is about the same in males and females. However, later in development, most of the female neurons die by apoptosis. These neurons can be rescued by the injection of testosterone. In contrast, many male bulbocavernosus neurons undergo apoptosis if an androgen receptor antagonist is given. Later in life, reduction in androgens leads to a reversible diminishing of bulbocavernosus dendritic arbors rather than to cell death.

Another example of sexually dimorphic neuronal circuits greatly affected by the sex hormones consist of the preoptic region (POA) of the hypothalamus and the bed region of the stria terminalis (BNST). These two areas have reciprocal axonal connections. A fetal surge in the level of testosterone promotes survival of the neurons in the male POA. In contrast, female POA neurons are gradually lost. There is evidence that the effect of testosterone in the male and female POA is not a direct one. Rather, there is much less estrogen in the female circulation and almost all is bound to alpha fetoprotein, which prevents it from reaching the brain. In the male POA, testosterone is converted to estrogen, which, by binding to estrogen receptors in the POA neurons, keeps them alive.

Another receptor, the glucocorticoid receptor (GR), is very abundant throughout the brain. The GR differs in one respect from the other steroid receptors by being in the cytoplasm until it binds its ligand. Then the GR enters the nucleus, dimerizes, and activates the genes associate with stress.

Stress responses in the brain and other organs are orchestrated by the hypothalamus—pituitary—adrenal cortical axis. Corticotropin-releasing hormone, made in the hypothalamus, as will be described further in the chapter, binds to receptors on the surface of the pituitary gland, causing them to release adrenocorticotropin (ACTH) into the bloodstream. ACTH in turn binds to its receptors on the plasma membrane of the adrenal cortex causing it to synthesize and release glucocorticoids into the bloodstream (Fig. 13.1). The glucocorticoids enter the brain and induce the stress response, often called the fight or flight response, by acting via the GR.

Glucocorticoids are essential during fetal brain development. In mice, maternal licking of the pups produces modifications of the GR gene. These modifications produce increased transcription of the GR mRNA, leading to the synthesis of more GR. Once the additional GR is synthesized and binds to glucocorticoids,

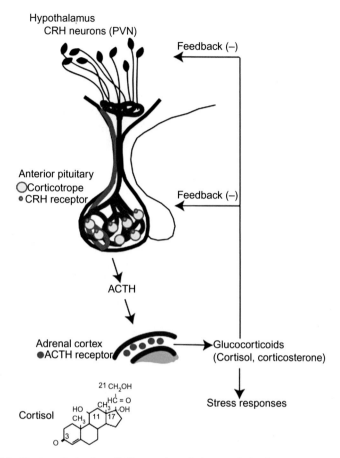

FIGURE 13.1 The hypothalamic—pituitary—adrenal stress axis is shown.

Hypothalamic CRH neurons in the paraventricular nucleus project their neuroterminals to the external zone of the median eminence. CRH travels through the portal vasculature to cause the synthesis and release of ACTH from corticotropes. This molecule acts at the adrenal cortex to cause the biosynthesis and release of glucocorticoids. These are released into general circulation to mediate the stress response, and they feedback in a negative manner at the corticotropes of the pituitary and the CRH neurons of the PVN (indicated by *long arrows*). The structure of cortisol, the major glucocorticoid released by the adrenal cortex in humans, is shown (bottom). *ACTH*, adrenocorticotropic hormone; *CRH*, corticotropic-releasing hormone; *PVN*, paraventricular nucleus.

Squire LR, Berg D, Bloom FB, Du Lac S, Ghosh A, Spitzer NC. Fundamental Neuroscience. 4th ed. Oxford: Elsevier; 2013 [Chapter 13]. Fig. 38.3, p. 803. With permission.

it causes decreased activation of genes coding for hypothalamic hormones, which will be discussed, and a more muted response to stress. Also as adults, these licked mice groom their offspring more, leading to a multigenerational blunting of stress-induced behaviors. Conversely animals deprived of licking and other tactile responses have hyperactive stress responses, leading to increased susceptibility to numerous diseases.

Two other steroid hormone systems are very important in CNS development. Vitamin A, which comes from fresh fruits and vegetables, is converted to retinoic acid. Retinoic acid and its receptor are very important in early patterning of the hindbrain. Also, retinoic acid is converted into retinol. Trans-retinol combines with the GPCR opsin to produce rhodopsin, which is the light-sensing molecule in the photoreceptor cells of the retina. Vitamin A deficiency is a major cause of childhood blindness in developing countries.

Vitamin D metabolites bind to the calcitriol receptor. This receptor forms a dimer with the retinoic acid receptor and turns on a number of important vitamin D metabolites, some of which are involved in brain development. One function controlled by vitamin D is calcium regulation. Another function is the regulation of nerve growth factor (NGF) formation. Too little vitamin D causes decreased NGF synthesis and increased neuronal apoptosis in the developing CNS. Vitamin D deficiency is associated with autism and schizophrenia as well as with age-related neurodegenerative diseases including Alzheimer's and Parkinson's diseases.

13.2 HYPOTHALAMUS

The hypothalamus consists of groups of neuropeptide synthesizing and secreting cells, which control a whole range of behavioral and metabolic processes through interactions with specific receptors on their target cells. There are a large number of small nuclei in the hypothalamus, some of which connect to various regions of the brain. Other nuclei secrete the neuropeptides that they synthesize directly into the blood. These are called neurohormones.

Many of the secreted neuropeptides affect human development including that of the brain. Their receptors are all members of the G-protein coupled receptor (GPCR) superfamily. The organization of the paraventricular nucleus, which synthesizes and secretes the neurohormones discussed here, is shown in Fig. 13.2.

The first two neurohormones to be isolated were the closely related nine amino acid–containing peptides vasopressin (VP) and oxytocin (OT). They are both produced as much larger polypeptides in the endoplasmic reticulum and are cleaved to the mature form in secretory vesicles. After their final cleavage, they are secreted into the bloodstream as classic neurohormones (Fig. 13.3).

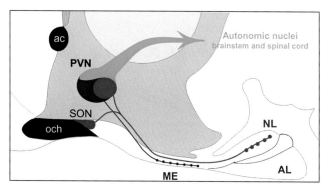

FIGURE 13.2 Functional associations of the paraventricular nucleus (PVH) with endocrine and autonomic systems are illustrated.

Axons of magnocellular neurons in the PVH and supraoptic nuclei (SON) (shown in *red*) traverse the internal layer of the median eminence (ME) and the infundibular stalk to terminate on a vascular plexus in the neural (posterior) lobe of the pituitary gland (NL). Parvicellular PVH neurons influence the functional activity of the anterior lobe of the pituitary gland (AL) through projections (shown in *blue*) that terminate on a fenestrated capillary plexus, the portal plexus, in the external zone of the median eminence. The terminals of these axons release peptides and neurotransmitters into the portal vasculature, which carries them to the anterior lobe of the pituitary. Parvicellular PVH neurons also give rise to descending projections (shown in *turquoise*) to autonomic nuclei in the brain stem and spinal cord. *ac*, anterior commissure; *och*, optic chiasm.

Squire LR, Berg D, Bloom FB, Du Lac S, Ghosh A, Spitzer NC. Fundamental Neuroscience. 4th ed. Oxford: Elsevier; 2013 [Chapter 13]. Fig. 33.5, p. 725. With permission.

VP controls two major physiological processes: constriction of blood vessels and water resorption by the kidney. If its regulation goes awry during pregnancy, the resulting problems can cause damage to the fetal brain.

There are a number of VP-producing neurons in other parts of the brain. VP receptors are also found in a number of brain regions. The production of VP in these regions can affect many different forms of behavior during development and in later life. Among them are social interactions, maternal bonding, sexual behavior, and response to stress.

OX controls uterine contraction during birth and subsequently milk ejection from the breast. It and its receptor are also found in various brain regions and are involved in a variety of behaviors during development and in adulthood. For example, there is an abundance of data indicating that early exposure to OX has important organizational affects on brain development. These include influencing many aspects of sexual and social behaviors.

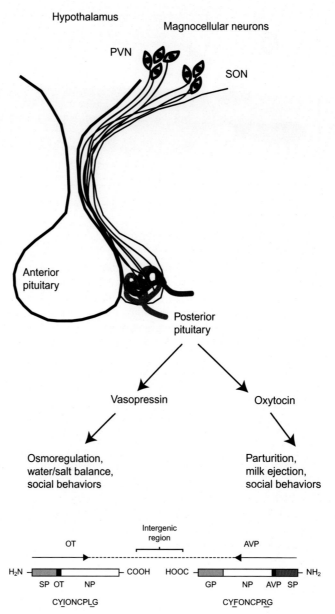

FIGURE 13.3 The hypothalamic—neurohypophysial control of vasopressin and oxytocin release is shown (top).

Large cells (magnocellular) in the paraventricular and supraoptic nuclei synthesize vasopressin or oxytocin, which are transported by axonal transport into nerve terminals located in the posterior pituitary gland (neurohypophysis). When stimulated by

(Continued)

In the 1950s Harris proposed that the secretion of hormones from the anterior pituitary gland is regulated by the release of factors by the hypothalamus into the hypophyseal portal veins. Many investigators attempted to isolate these substances with no success. Finally, in the early 1970s, Schally and Guillemin and Vale independently isolated several releasing factors and inhibiting factors from the hypothalamus. All of these are small peptides. For their discoveries, Guillemin and Schally shared the Nobel Prize in Physiology and Medicine in 1977.

One releasing factor is thyrotropin-releasing hormone, which causes thyrotropin release from the pituitary. This in turn produces thyroid hormone synthesis and release from the thyroid gland. Thyroid hormone, acting through its GPCR signal, is the master metabolic regulator in the body (Fig. 13.4).

Another releasing factor is corticotropin-releasing hormone, which stimulates ACTH release. ACTH through binding to its GPCR stimulates the synthesis and release of glucocorticoids from the adrenal cortex. As discussed previously, glucocorticoids are the major stress hormones in the body and essential for appropriate brain development (see Fig. 13.1).

A third releasing factor is growth hormone–releasing hormone, which stimulates growth hormone release. Growth hormone, through its GPCR, controls the development of all organs including the brain (Fig. 13.5).

The hypothalamus also secretes two inhibiting factors. One is prolactin release inhibiting factor; the other is growth hormone inhibiting factor, also called somatostatin. Many neurons in the brain also synthesize and secrete one of the releasing or release inhibiting factors together with a classic neurotransmitter. For example, somatostatin secreting, inhibitory neurons in various regions of the brain play key roles in brain development. This is likely to be the case with other releasing factor secreting neurons as well.

◄ depolarization, vasopressin and/or oxytocin are released from the posterior pituitary into general circulation. The structural organization of the vasopressin and oxytocin genes are shown (bottom). The RNA sequences are transcribed off of a single gene but in opposite directions, indicated by arrows. Each gene is made up of three exons that encode a signal peptide, the neurohormone (either vasopressin or oxytocin), and a neurophysin that may be involved in chaperoning the neuropeptides. Vasopressin also contains an additional sequence for a glycoprotein. The regions are separated by an intergenic sequence. The nonapeptide sequence of each neurohormone is shown below its respective genes, with differences indicated by underlines. *AVP*, arginine vasopressin; *GP*, glycoprotein; *NP*, neurophysin; *OT*, oxytocin; *PVN*, paraventricular nucleus; *SON*, supraoptic nucleus; *SP*, signal peptide.

Squire LR, Berg D, Bloom FB, Du Lac S, Ghosh A, Spitzer NC. Fundamental Neuroscience. 4th ed. Oxford: Elsevier; 2013 [Chapter 13]. Fig. 38.8, p. 814. With permission.

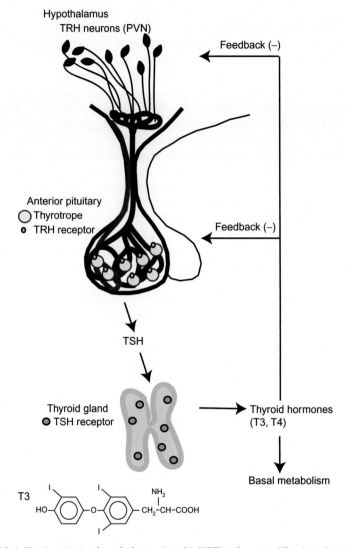

FIGURE 13.4 The hypothalamic−pituitary−thyroid (HPT) axis controlling basal metabolic functions is shown.

TRH neurons in the paraventricular nucleus of the hypothalamus project their neuroterminals to the external zone of the median eminence. TRH travels through the portal vasculature to cause the synthesis and release of TSH from thyrotropes. This molecule acts at the thyroid gland to cause the biosynthesis and release of the thyroid hormones, T3 and T4. These are released into the general circulation to affect metabolism. Negative feedback of T3 and T4 at the thyrotropes of the pituitary and at the TRH neurons of the PVN is indicated by the long arrows. The structure of T3 is shown (bottom). *PVN*, paraventricular nucleus; *T3*, triiodothyronine; *T4*, tetraiodothyronine or thyroxine; *TRH*, thyrotropin-releasing hormone; *TSH*, thyroid-stimulating hormone.

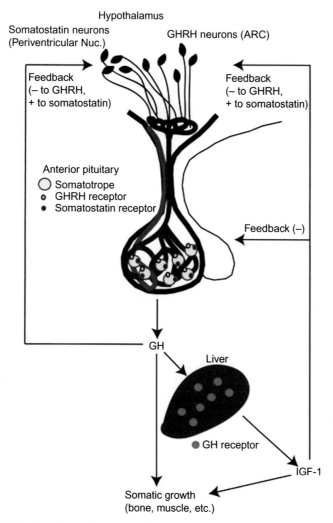

Hypothalamus

Somatostatin neurons
(Periventricular Nuc.)

GHRH neurons (ARC)

Feedback
(– to GHRH,
+ to somatostatin)

Feedback
(– to GHRH,
+ to somatostatin)

Anterior pituitary
○ Somatotrope
◎ GHRH receptor
● Somatostatin receptor

Feedback (–)

GH

Liver

● GH receptor

IGF-1

Somatic growth
(bone, muscle, etc.)

FIGURE 13.5 The hypothalamic control of somatic growth is shown.

GHRH neurons in the arcuate nucleus, and somatostatin neurons in the periventricular region of the anterior hypothalamus, provide the stimulatory and inhibitory inputs to pituitary somatotropes, respectively (via the portal capillary vasculature). The synthesis and release of GH is stimulated (GHRH) or inhibited (somatostatin) at the somatotropes. This molecule acts at the liver to cause the biosynthesis and release of IGF-I. Both GH and IGF-I are released into the general circulation to affect somatic growth, and they both feedback negatively at the somatotropes of the pituitary and the GHRH neurons of the arcuate nucleus. IGF-I and growth hormone also feedback to stimulate somatostatin. *ARC,* arcuate nucleus; *GH,* growth hormone; *GHRH,* growth hormone-releasing hormone; *IGF-I,* insulin-like growth factor I.

*From Squire LR, Berg D, Bloom FB, Du Lac S, Ghosh A, Spitzer NC. Fundamental Neuroscience. 4th ed.
Oxford: Elsevier; 2013 [Chapter 13]. Fig. 38.5, p. 807. With permission.*

GENERAL REFERENCES

Book Chapters

Kandel ER, Schwartz JH, Jessel TM, Siegelbaum SA, Hudspeth AJ. *Principles of Neural Science*. 5th ed. New York: McGraw Hill; 2013 [Chapter 47].

Kreier F, Swaab DF. *Handbook of Clinical Neurology*. Vol. 95. Oxford, UK: Elsevier [Chapter 23]; 2009. Available from: https://www.sciencedirect.com/handbook/handbook-of-clinical-neurology/vol/95/suppl/C.

Squire LR, Berg D, Bloom FB, Du Lac S, Ghosh A, Spitzer NC. *Fundamental Neuroscience*. 4th ed. Oxford, UK: Elsevier; 2013 [Chapter 13].

Review Articles

Guillemin R. Neuroendocrinology: a short historical review. *Ann N Y Acad Sci*. 2011;1220:1−5.

Receptors and the development of the enteric nervous system

14

CHAPTER OUTLINE

14.1 THE ENTERIC NERVOUS SYSTEM

The enteric nervous system (ENT) in humans is an interlocking system of approximately 50 million neurons and the same number of glia that form a web surrounding the whole length of the gut from esophagus to anal sphincter (Fig. 14.1). This system is sometimes referred to as the second brain because it communicates with the brain via the vagus nerve. However, it can function independently to control gut peristalsis and the secretion of a variety of enzymes and hormones in the absence of communication with the brain.

The enteric neural cells derive from the neural crest during late embryogenesis and migrate, in some cases more than 6 inches, to reach their final destinations. The enteric neural cell migration occurs soon after birth when the infant begins eating. During their migration, trophic factors released in some cases by the gut cells that they will innervate play a major role. The most important of these trophic factor−receptor systems is the glial cell line−derived growth factor (GLDF) produced by the targeted gut endothelial cells with its receptor, RET. GLDF attracts the RET receptor, a member of the tyrosine kinase superfamily, on the migrating nerve cells. Studies from a number of laboratories have shown that this system is necessary for the differentiation, survival, and migration of the enteric nerve cells. Babies born without this system suffer from Hirschsprung disease, that is, the absence of the ENS from the large intestine. Affected individuals cannot digest solid food and require surgery to remove the noninnervated bowel.

Other trophic systems similar to those employed during brain development are also utilized by the ENS. These include NT3 and its receptor, TrkC, as well as endothelin-3 and the endothelin B receptor. Another system involved in the

Receptors in the Evolution and Development of the Brain. DOI: https://doi.org/10.1016/B978-0-12-811012-6.00014-5

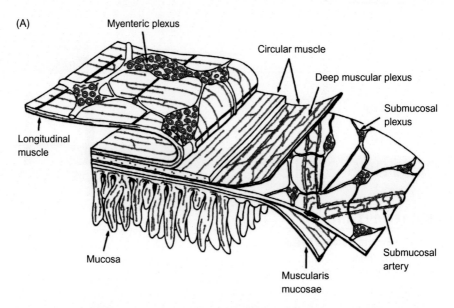

(A)

Myenteric plexus

Circular muscle

Deep muscular plexus

Submucosal
plexus

Longitudinal
muscle

Mucosa

Muscularis
mucosae

Submucosal
artery

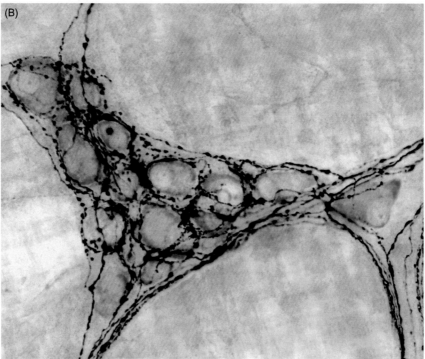

(B)

FIGURE 14.1

(A) Enteric neurons form two conspicuous and extensive networks, or plexuses, of ganglia
and connectives located between the longitudinal and circular muscle layers of the wall (the
myenteric plexus) and the circular muscle layer and the inner mucosal layer (the
submucosal plexus). The deep muscular plexus, periglandular plexus, and villous plexus are
additional networks of enteric neurons and their processes. (B) Autonomic preganglionic

(*Continued*)

migration is sonic hedgehog, produced by epithelial cells, which binds to the Patched receptor expressed by enteric neurons. Finally bone morphogenetic proteins (BMPs) and their receptors, the BMPRs 1,2, and 3, are important in enteric neural cell migration. Developing and migrating enteric neurons also secrete neurotransmitters including Glu, GABA, serotonin, and dopamine that can function as tropic, trophic, and/or guidance molecules via binding to their respective receptors.

The enteric neurons migrate as chains linked together by adhesive proteins acting as receptors. There is not much known about the attractive and repulsive molecules used during enteric nerve cell migration, the exception being the Netrins and their receptors, which are described in Chapter 12, key receptors involved in laminar and terminal specification and synapse construction.

One major difference from brain development is that there is almost no apoptosis during ENS development. A reasonable explanation, based on the theoretical arguments proposed in Chapter 7, A testable theory to explain how the wiring of the brain occurs, is that because the human ENT contains roughly 1/200th the number of neurons as the human brain, the ENT does not require a selection mechanism to make the correct synaptic connections.

14.2 THE MICROBIOTA

The colonization and growth of the gut microbiota occurs in coordination with the development of the ENT. The mature microbiota in the human body consists of millions of different species of protists (mainly bacteria), together with viruses that live in a symbiotic relationship in these protists. The genomes of the species contained in the microbiota are collectively called the microbiome. Approximately half of these species are found in the gut. There are approximately a trillion protists in the healthy digestive tract, more than all the cells of human origin in the body. Until quite recently the number of species of bacteria and other protists inhabiting the gut was unknown. However, with the development of sensitive, rapid, and inexpensive DNA sequencers, the number of species identified in the intestinal tract has grown exponentially.

The initial colonization of the digestive tract by protists occurs as the vaginally delivered baby passes through the nonsterile environment of the growth

◄ terminals innervating postganglionic neurons. This example of vagal preganglionic projections (*brown*, labeled with the tracer *Phaseolus vulgaris*) innervating myenteric ganglion neurons (*blue*, stained with Cuprolinic blue) in the stomach wall illustrates that autonomic preganglionic axons can form extensive networks controlling postganglionics.

Squire LR, Berg D, Bloom FB, Du Lac S, Ghosh A, Spitzer NC. Fundamental Neuroscience. 3rd ed. Oxford: Elsevier; 2008 [Chapter 34]. Fig. 34.9, p. 737. With permission

canal, through the vagina, and into the environment. It has been demonstrated that the major bacterial species in the neonatal intestinal tract of a vaginally delivered baby correspond to those in the mother's vagina. In contrast, the digestive tracts of babies delivered by Caesarean section contain mainly bacterial and other protist species found in the skin. There is now evidence that vaginally delivered babies show higher cognitive abilities that those delivered by Caesarean sections. Also there is recent evidence that this difference is in part due to the differences in microbiota species between the two groups.

Another source of difference in the microbiota is whether the baby is breast or bottle fed. Again there is evidence suggesting that breast-fed babies do better on a group of tests measuring cognition.

The organisms making up the gut microbiota influence the developing brain in a number of ways, some of which involve the production of substances that act as neurotransmitters and neuropeptides, which directly affect the central nervous system (CNS). Also the microbiota influences the ENT, which in turn influences the brain (Fig. 14.2).

It is now known that the largest quantity of neurotransmitters found in the body is in the intestinal microbiota, not in the brain. Both during and after development, these neurotransmitters bind to their respective receptors in the ENT potentially influencing the brain. The microbiota also produces a large quantity of small chain fatty acids that enter the bloodstream and cross the blood–brain barrier due to their hydrophobicity. These fatty acids serve as an important energy source for the developing brain. They also help to maintain the barrier properties of the blood – brain barrier (Fig. 14.3).

The best evidence for the importance of the microbiota in brain development comes from studies with mice conceived, born, and raised in a sterile environment (GF mice). These animals show altered behavioral and cognitive changes during development. These changes have been associated with lowered levels of BDNF and several neurotransmitters in the brain. These include dopamine, GABA, and serotonin. Also GF mice have higher levels of ACTH, a key stress hormone, than do normal animals. This increase causes an elevated response to restraint stress, which can be partially reversed by recolonization of the intestinal microbiota by a fecal transplant containing the microbiota of a normal mouse.

A recent publication demonstrates the role of the microbiota in a developmental disease of the brain. Some babies are born with small or large bubbles in the veins of their brain. If these bubbles burst they can cause microhemorrhages, leading to seizures, strokes, or even death. In a model of this disease in mice, it has been discovered that pups given antibiotics followed by fecal transplants do not get bubbles in their brains. It is thought that a lipopolysaccharide produced by a Gram-negative bacterial species is responsible for bubble formation. By removing that particular bacterial species through antibiotic treatment followed by a fecal transplant, one removes the source of the bubble-forming molecules.

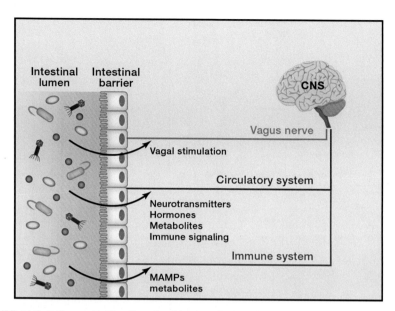

FIGURE 14.2 Pathways linking the microbiome and central nervous system.

Signals from the intestinal microbiome may potentially traffic to the central nervous system (CNS) via several mechanisms. (1) Direct activation of the vagus nerve from the enteric nervous system to the CNS. (2) Production of, or induction of, various metabolites that pass through the intestinal barrier and into the circulatory system, where they may cross the blood–brain barrier to regulate neurological function. (3) Microbial associated molecular patterns (MAMPs, such as LPS, BLP, and PSA) and metabolites produced by the microbiome can signal to the immune system. Immune cells (and particularly their cytokines) can influence neurophysiology.

Sampson TR, Mazmanian SK. Control of brain development, function, and behavior by the microbiome. Cell Host Microbe. *2015;17:565–576. PMID: 25974299. Fig. 1. With Permission.*

There is also suggestive evidence that autism spectrum disorder (ASD), which will be discussed in Chapter 17, Role of trophic factors and receptors in developmental brain disorders, may be associated with an abnormal microbiota. Animal models of this disease can be improved by altering the intestinal microbiota. Also depression in humans and in animal models can also be positively affected by an altered microbiota.

Research into the influence of the microbiota on the development of the gut, brain, and other organs is in its relative infancy. However, one may wonder whether the increased use of antibiotics, which can drastically alter the intestinal microbiota, can be at least partly responsible for the great increase in ASD and also attention deficit disorder in developed societies.

One area in which studies of the human microbiota has been applied widely in clinical medicine is the treatment of *Clostridium difficile* infections in the ENT after antibiotic use. These bacteria secrete toxic molecules that bind to receptors on the

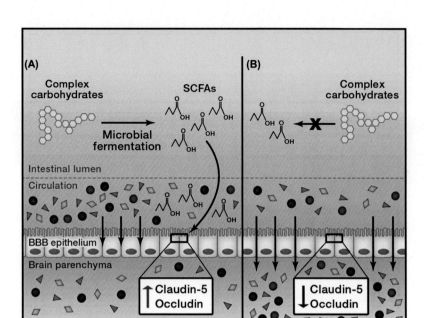

FIGURE 14.3 Microbiome influence on blood—brain barrier integrity.
Intestinal microbes are capable of fermenting complex carbohydrates into short chain fatty acids (SCFAs). (A) Microbially produced SCFAs signal to epithelial cells that create the blood—brain barrier (BBB) and increase production of the tight junction proteins claudin-5 and occludin. This leads to a tight and selective barrier, preventing undesired metabolites from entering the brain parenchyma. (B) In the absence of microbial fermentation, no SCFA signaling occurs, and tight junction proteins are repressed. This leads to increased permeability of the BBB, and a loss of the selective barrier to serum metabolites.

Sampson TR, Mazmanian SK. Control of brain development, function, and behavior by the microbiome. Cell Host Microbe. 2015;17:565—576. PMID: 25974299. Fig. 2. With Permission.

surface of the intestine, producing severe damage to the cells of the ENT. In cases in which treatment with additional antibiotics for *C. difficile* infections has proved ineffective, treatment of patients with fecal transplants consisting of the microbiota of normal individuals has proven to be extremely effective.

14.3 HARMFUL VIRUSES THAT AFFECT THE DEVELOPMENT OF THE BRAIN

A number of viruses can infect the developing CNS and all use receptors on the neural cell surface to gain entry to the host cells. This is essential to the survival and reproduction of all viruses since they cannot replicate outside the host cell. We will discuss several viruses as examples of causative agents for potentially severe diseases that affect the developing brain. Poliovirus is a single-stranded positive RNA-containing virus. This enables the RNA to be used to synthesize proteins. Those individuals who were born before the late 1950s when the first polio vaccine became available remember the widespread fear of damage to the CNS produced in children and young adults by infection with the virus. The key to the production of this vaccine was the cultivation of human cells in cell culture, which allowed the isolation and purification of enough virus particles to inject into many people thereby producing immunity. The isolation and purification of poliovirus was accomplished by John Enders, Thomas Weller, and Fredrick Robbins. For this accomplishment they received the Nobel Prize in Physiology and Medicine in 1954.

Polio is a member of the enterovirus family, which enters the host individual through oral delivery. Humans are the only natural host. Once in the gut, the virus binds to a receptor called CD155, which is found in many cells in the body, and whose known function is to establish and maintain adherens junctions between epithelial cells in the intestinal tract and elsewhere. After poliovirus binds to this receptor, it enters the cell in clathrin-coated vesicles. It is then released from endosomes into the cytoplasm and replicates. The newly synthesized viral particles kill the infected cell and enter other cells including skeletal muscle. In most cases polio produces mild symptoms, but in approximately 1% of individuals the virus enters motor neurons and produces an infection causing paralysis. In cases in which motor neurons innervating the respiratory neurons are infected, breathing difficulty ensues that can lead to death.

Poliovirus infects the CNS opportunistically. The CNS is not important in the viral life cycle; rather it is a dead end. Researchers do not understand why the virus most commonly causes damage to motor neurons in infants and young individuals. It is possible that they are less likely to have natural immunity than are older individuals.

Quite recently, a novel treatment employing a hybrid poliovirus/rhinovirus has targeted a deadly brain cancer, glioblastoma, with some success. The virus recognizes receptors found only on the glioblastoma cells and infects them. This leads immune cells in the brain to attack the infected cancer cells, destroying them.

The herpes simplex virus (HSV) infects individuals mainly through sexual contact. It is an enveloped virus that contains a double-stranded DNA surrounded by a protein capsid. Surrounding the capsid is a membrane that contains phospholipids derived from the host cell and integral membrane proteins. These proteins are translated from mRNAs transcribed by the viral DNA. The viral proteins bind to specific proteoglycans on the host cell and then to their receptor, nectin-1. This binding stimulates the fusion of the virus and host membranes and release of the viral capsid into the host cell cytoplasm. The virus genome is quite large, containing genes necessary for viral replication and capsid protein synthesis. Some HSV genes are also transcribed into mRNAs, which are translated into membrane proteins capable of insertion into the host plasma membrane. This in turn allows the virus to bud from the host cell surface into the cytoplasm, where it replicates, killing the host cell and infecting other cells.

Over 50% of the population of the United States is infected with HSV and the virus remains in CNS neurons in an inactive state. In rare cases it is activated, enters the 5th cranial nerve, and is transported to the brain. Once in the brain, HSV produces a viral encephalopathy mainly in the temporal lobe. The infection is treated with high doses of an antiviral agent acyclovir. However, the infection still is fatal in 30% of its victims. Also survivors have a variety of brain damage. Again, the encephalopathy is not a necessary part of the HSV life cycle.

Another virus that has been in the news for several years is the Zika virus. It is a member of the Flavivirus family, members of which are yellow fever, Dengue, and West Nile viruses. All of these viruses are transmitted to humans by the bite of a mosquito. In the case of Zika, most infections are caused by the bite of *Aedes aegypti*. Infection resulting from sexual contact may also occur.

Zika is a positive-strand RNA virus. Its RNA is surrounded by a capsid and a membrane that contains proteins translated from the Zika virus mRNA. These membrane proteins bind to AXL, an integral membrane tyrosine kinase. AXL is important to the function of the immune system and to maintain the blood−brain barrier. Normally Zika produces a mild infection. However, in pregnant women, especially during the first trimester, the virus can cross the placental blood−fetal barrier and enter the brain. After entry, Zika preferentially infects radial glial cells (RGCs) because they contain a high concentration of AXL on their plasma membranes. The virus kills these cells, which are necessary for the differentiation of cerebral cortical neurons, astrocytes, and oligodendrocytes, as discussed in Chapter 8, The development of the cerebral cortex. The loss of RGCs results in an abnormally small brain, as well as to other irreversible brain damage. In some cases the fetuses die, either before or shortly after birth. In other cases, the children survive but with significant brain damage. The Zika brain infection is opportunistic since it is not essential to the virus's reproductive cycle.

At this time there is no Zika vaccine and the only protection for women who are or hope to become pregnant is to avoid areas that harbor infected mosquitos. Hopefully this situation will be changed in the near future by the development of a vaccine.

There are many other viruses that can infect the developing nervous system. One of these is the rabies virus, which is a negative single-stranded RNA virus. Its RNA is surrounded by a cylindrical capsid and a membrane containing a viral coated glycoprotein (G) protein. As opposed to the viruses discussed previously, rabies has a large host range and can infect all mammals. Also rabies must enter the nervous system for successful replication and subsequent transmission to occur. It therefore evokes significant behavioral changes in the host that lead to serial infections. Even though a vaccine has been available in the developed nations for decades, approximately 30,000 people still succumb to rabies infection yearly.

While all of the receptors for rabies have not been definitively identified, one is the p75 neurotrophin receptor. On viral entry through a puncture wound produced by a rabid animal, the viral surface G protein binds to one of its cell surface receptors, and is endocytosed into dendrites of neurons in the peripheral nervous system. The virus is then transported to the cell body by retrograde axonal transport through a dynein-mediated mechanism. Upon reaching axonal terminals that connect to the CNS, the virus is secreted. Again it enters motor and sensory neurons and is transported into the brain by dynein.

Once in the brain, which can take between a few days and months, depending upon the distance the virus must be transported, rabies virus produces a generalized inflammation and encephalitis. It also produces characteristic symptoms including hydrophobia, the fear of water, which is caused by the inability to swallow. This allows the virus to rapidly multiply in salivary glands, which are infected via the nerve cells innervating them. Dogs and other mammals become very aggressive, biting other individuals and spreading the disease.

REFERENCES

Recommended Book Chapters

Kandel ER, Schwartz JH, Jessel TM, Siegelbaum SA, Hudspeth AJ. *Principles of Neural Science*. 5th ed. New York: McGraw Hill; 2013 [Chapter 47].

Squire LR, Berg D, Bloom FB, Du Lac S, Ghosh A, Spitzer NC. *Fundamental Neuroscience*. 4th ed. Oxford, UK: Elsevier; 2013 [Chapter 34].

Strauss E, Strauss J. *Viruses and Human Disease*. 2nd ed. Oxford, UK: Elsevier; 2007.

Review Article

Sampson TR, Mazmanian SK. Control of brain development, function, and behavior by the microbiome. *Cell Host Microbe*. 2015;17:565−576.

Synaptic pruning and trophic factor interactions during development

15

CHAPTER OUTLINE

During development it is essential that unnecessary synapses be removed to produce a properly wired circuit. This process is called synaptic pruning. Inactive synapses are marked by the addition of two proteins to their surface, C1 and C3. Microglia recognize these synapses, phagocytose them, and digest them in lysosomes. Some neurons that lose synaptic trophic support don't die. Corollary 1, presented in Chapter 14, The roles of receptors and trophic factors in the development of the enteric nervous system and in the connections between viruses, the microbiota, ENT, and brain, provides a plausible explanation. For example, if a neuron contains three quanta of a receptor it will require three quanta of the complementary trophic factor. However, it is conceivable that the neuron can stay alive indefinitely by upregulating its receptors. This mechanism allows the neuron to produce a new neurite, which can locate another neurite that secretes a complementary trophic factor, and form a synapse with it. A mechanism similar to synaptic pruning occurs in the mature brain and is called synaptic plasticity.

Trophic factors and receptors play major roles in deciding which neurons will live and which will die, and in some cases the decisions are final. As first mentioned in Chapter 7, A testable theory to explain how the wiring of the brain occurs, another major decision, referred to as synaptic pruning, is also dependent on trophic factor–receptor interactions. These interactions are of great importance in determining the correct circuitry of the mature brain. Trillions of synaptic connections form during human brain development, and many of these disappear during development or afterward. These connections are in some cases replaced by others.

Early in brain development, waves of spontaneous electrical activity flow through the brain, strengthening some synapses and weakening others. Two proteins, C1 and C3, associated with the classical complement cascade in the blood, play key roles in synaptic pruning. These proteins, which are synthesized by developing neurons in response to astrocytes' signals, are transported to the surface of synapses. Microglial cells recognize C1 and C3, which can be thought of as receptors, by binding to the synaptic surface. This specific binding results in the destruction of that synapse. Mice deficient in the C1 and/or C3 protein contain too many synapses (Fig. 15.1).

Receptors in the Evolution and Development of the Brain. DOI: https://doi.org/10.1016/B978-0-12-811012-6.00015-7

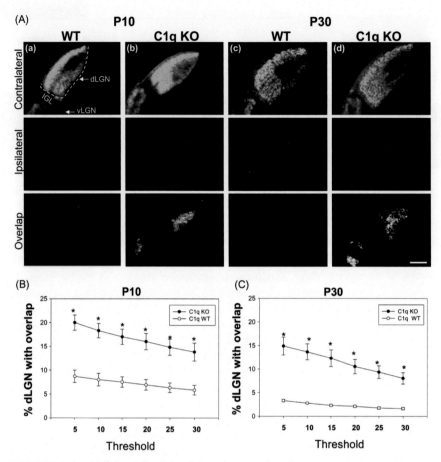

FIGURE 15.1 C1q-deficient mice have defects in synaptic refinement and eye-specific segregation.

(A) Retinogeniculate projection patterns visualized after injecting β-cholera toxin conjugated to Alexa 594 (CTβ-594) dye (*red*) and CTβ-488 (*green*) into left and right eyes of WT and C1q KO mice. C1q KO mice at P10 (Ab) and P30 (Ad) have significant intermingling (*yellow*, overlap) between RGC axons from left and right eyes compared with littermate WT controls (Aa and Ac). Scale bar, 200 μm. (B and C) Quantification of the percent of dLGN-receiving overlapping inputs in C1q KO versus WT controls at P10 (B) and P30 (C). C1q KO mice exhibit significantly more overlap than WT mice, regardless of threshold. Data are represented as mean ± SEM [P10, $n = 8$ mice, $P < .01$ (*t*-test) at 5% threshold and $P < .05$ at 30% threshold; P30, $n = 6$ mice, $P < .009$ at 5% threshold, and $P < .01$ at 30% threshold].

Stevens B, Allen NJ, Christopherson KS, Vazquez LE, Nouri N, Howell GR, et al. Cell. 2007;131:1164–1178. PMID: 18083105. Fig. 4. With permission.

A good question that can be raised here is: what factor or factors tip the balance between a particular neuron losing some nerve endings and therefore some trophic support, but surviving, and the same neuron becoming trophically deprived of enough trophic support to initiate apoptosis? At present, there is no obvious answer. However, one possible explanation follows from the hypothesis concerning the quantization of trophic factor requirements discussed in Chapter 7, A testable theory to explain how the wiring of the brain occurs. Suppose that a neuron loses some quantal trophic support as one or more synapses are pruned. However, if the neuron still maintains a sufficient number of synapses to fulfill its minimal quantal requirements, it survives. If not, it dies. One can imagine that at an inactive postsynaptic ending, the dendrite or axon deprived of its trophic support secretes more trophic factor molecules. This in turn attracts another axon or dendrite, which can supply the necessary trophic support (Fig. 15.2).

During development, formation of synaptic connections and their stabilization is mediated to a large degree by trophic factor–receptor interactions occurring between the pre- and postsynaptic endings. The activity of a particular synapse, that is, the rate of neurotransmitter release compared with that of adjacent synapses, is an important factor as well. Needless to point out, trophic factors are released from presynaptic endings when neurotransmitter release occurs, and in many cases the neurotransmitters are themselves trophic factors. There is also evidence that trophic factors including BDNF and endocannabinoids can be released from postsynaptic endings as well, as was discussed in Chapter 10, The key roles of BDNF and endocannabinoids at various stages of brain development including neuronal commitment, migration, and synaptogenesis.

Many reports indicate that newly formed synapses that are active are likely to be stable, while those that aren't disappear. One can postulate that the reason for the disappearance of inactive synapses is caused, at least in part, by the lack of trophic support. In many developing neural systems, activity is stimulated by various experiences. A classic example of this occurs in the formation of columns in the visual cortex, as was discussed in Chapter 10, The key roles of BDNF and endocannabinoids at various stages of brain development including neuronal commitment, migration, and synaptogenesis. These columns result from the convergence of afferent axons from the same region of the retina onto a single cortical neuron. Their synchronous firing, and therefore release of trophic factors, also increases the strength of synapses on surrounding neurons. Synapses from the other eye are concomitantly inactivated, leading to their disappearance. Ultimately this process leads to the formation of a column. Columns from one retina alternate with those from the other (Fig. 15.3). However, If one eye is closed during the so-called critical period of development, the axons emanating from the retinal ganglion cells of the open eye will form synapses with the cortical neurons from the closed eye. Meanwhile any synapses formed by axons from the closed eye disappear, presumably because they lack suitable trophic factor support caused by decreased activity. (See Fig. 10.2).

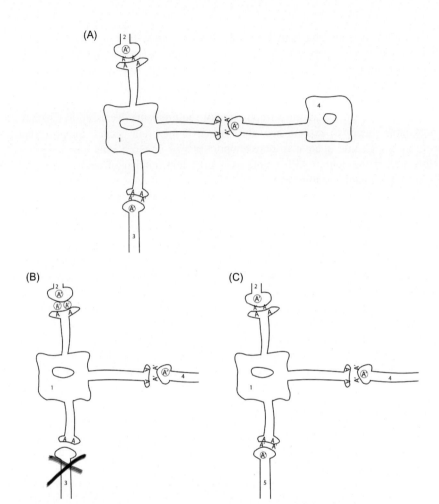

FIGURE 15.2 A cartoon illustrating synaptic pruning during development or synaptic plasticity in the adult.

(A) Shows cell 1, which requires 3 quanta of A′ contributed by the synapses formed on cell 1 by processes (either axons or dendrites) from cells 2, 3, and 4. (B) Shows that cell 1 loses 1 quanta of trophic support caused by the death of cell 3 or the retraction of its process. However, it still receives 2 quanta of A′ from cells 2 and 4. enabling it to survive. (C) Cell 1 forms a synapse with cell 5 and regains 3 quanta of A′. (This cartoon should not imply that cell 1 requires only quantal support by A′. It almost certainly requires trophic support from other trophic molecules as well.)

Adapted from Fine R, Rubin J. J Am Geriatr Soc. 1988;36:457–466. PMID: 2834427. Fig. 1. With permission.

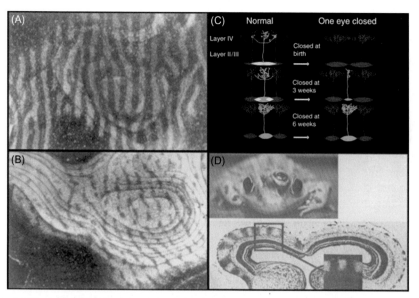

FIGURE 15.3 Synapse elimination in the visual system.

(A) Ocular dominance columns of the neonatal monkey primary visual cortex in layer IVC, revealed by injecting [3H]-proline into the vitreous of one eye. Light stripes (columns) represent sites containing the anterograde transported 3H-amino acid from the injected eye. Dark regions are occupied by axons driven by the other eye. (B) Monocular deprivation by lid suture of one eye (2 weeks after birth for a period of 18 months) resulted in the shrinkage of the columns representing the deprived eye (*dark stripes*) and an expansion of the columns of the nondeprived eye (*light stripes*). (C) A schematic representation of ocular dominance column development represents the way in which a gradual segregation ocular dominance column could lead to the end of the critical period and progressively more modest effects of monocular deprivation as development ensues. (Top) At birth, the afferents from the two eyes (*red and green ovals*) overlap completely in layer IV, and thus each eye is capable of maintaining inputs everywhere. At this young age, monocular deprivation would allow the nondeprived eye to remain in all parts of layer IV so that the entire cortex would be dominated by the red inputs. (Middle) In nondeprived animals, the two sets of afferents become progressively more segregated with age, meaning that by three weeks there would be regions of layer IV that are exclusively driven by the red or, as shown, green afferents. Once an eye's inputs are removed from a territory, it can no longer reoccupy that territory when the other eye is silenced. Hence monocular deprivation (that begins at 3 weeks) will spare a small strip of the inactive eye's territory (in this case the *green regions*). (Bottom) Once segregation is complete then monocular deprivation has no effect and the critical period is over. (D) A remarkable example showing how interactions between two eyes can cause segregation was found in frogs in which a third eye is implanted (at the tadpole stage) next to one of the eyes and projects with the native eye to the same optic tectum (ordinarily each frog tectum is monocular). After injection of H3-proline into the normal eye, one optic tectum of a three-eyed frog shows dark and light bands strikingly similar to the ocular dominance columns observed in monkeys (D).

Squire LR, Berg D, Bloom FB, Du Lac S, Ghosh A, Spitzer NC. Fundamental Neuroscience. 4th ed. Vol. 18. Oxford, UK: Elsevier; 2013 [Chapter 10]. Fig. 19.15, p. 450. With permission.

As was discussed in Chapter 10, The key roles of BDNF and endocannabinoids at various stages of brain development including neuronal commitment, migration, and synaptogenesis, BDNF plays a key role in this process. A similar mechanism of column formation in other cortical areas including the motor and somatosensory cortices has also been demonstrated.

Critical periods for other cognitive and behavior changes occur during human development. We are all familiar with the fact that it is very easy to master a second language before the age of 5 but much harder thereafter.

The role of a critical period in human psychosocial development is also demonstrated by a study comparing two groups of infants. One half of the children were raised in a nursing home attached to a women's prison with access to their mothers; the other half in an orphanage. The latter group of children developed a range of pathological behaviors including depression, withdrawal, and susceptibility to infections, compared with children raised with maternal access.

The replacement of synapses after development is completed is called neuronal plasticity. It can result in remarkable connectional rearrangements after damage to a particular brain region. An example of this plasticity is seen in the somatosensory cortex. Normally, there is a topological map corresponding to the various regions of the body, for example, fingers, hands, etc. In case there is the loss of a finger, the cortical connections in the portion of the somatosensory map corresponding to the amputated finger diminish, allowing those in the adjacent territory representing adjacent fingers to expand to fill the vacant region. Also, victims of serious strokes that cause a major loss of a specific cognitive function can make remarkable recoveries of function by redeploying already present neurons and expanding their synaptic targets. These rearrangements can at least partially be understood employing the trophic mechanism proposed in Fig. 15.2.

GENERAL REFERENCES

Recommended Books

Kandel ER, Schwartz JH, Jessel TM, Siegelbaum SA, Hudspeth AJ. *Principles of Neural Science*. 5th ed. New York: McGraw Hill; 2013 [Chapters 12, 62, 67].

Squire LR, Berg D, Bloom FB, Du Lac S, Ghosh A, Spitzer NC. 4th ed. *Fundamental Neuroscience*. vol. 18. Oxford, UK: Elsevier; 2013 [Chapter 10].

Review Articles

Stephan AH, Barres BA, Stevens B. The complement system: an unexpected role in synaptic pruning during development and disease. *Annu Rev Neurosci*. 2012;35:369−389.

Dong Y, Taylor JR, Wolf ME, Shaham Y. Circuit and synaptic plasticity mechanisms of drug relapse. *J Neurosci*. 2017;37:10867−10876.

Receptor-mediated mechanisms for drugs of abuse and brain development

16

CHAPTER OUTLINE

This chapter will be devoted to a discussion of potentially addictive drugs and their effects on brain development. Marijuana has already been discussed since there is much evidence for its beneficial effects as well as for its harmful ones. Individuals who use marijuana suffer from the characteristic changes as the drugs discussed in this chapter, including physical dependency, tolerance, and difficulty in withdrawal. These changes, however, do not appear to be as powerful, and not do not lead to addiction in most cases.

Over the last several decades, it has become clear that all potentially addictive drugs work through the same receptor mechanism, mainly in an area of the brain called the nucleus accumbens, which is referred to as the pleasure center. Neurons carrying dopamine from the ventral segmental area interact with the excitatory glutamate (Glu) neurons from the cerebral cortex, causing dopamine release from dopaminergic neurons in the nucleus accumbens. which evokes a pleasurable sensation. Pathways from the amygdala to the nucleus accumbens also play an important role (Fig. 16.1). While all potentially addictive drugs use this mechanism, many have other independent actions as well, involving other receptors and transmitters and/or neuropeptides.

Cocaine is derived from the leaves of the coca plant and its leaves have been chewed by inhabitants of the Andes for thousands of years to help adjust to life at high altitude. Cocaine causes a feeling of hyperalertness along with a strong rush of pleasurable emotions. The hyperalertness allows the user to stay awake for many hours without drowsiness.

Cocaine is a small, very hydrophobic molecule and can very readily cross the blood–brain barrier and enter the brain. Once in the brain, its major mechanism of action is to bind to and inhibit the dopamine transporters that pump released dopamine back into the presynaptic cells, thereby ending its action. However, in the presence of cocaine, the postsynaptic dopaminergic receptors are able to bind

Receptors in the Evolution and Development of the Brain. DOI: https://doi.org/10.1016/B978-0-12-811012-6.00016-9

Neurochemical neurocircuits in drug reward

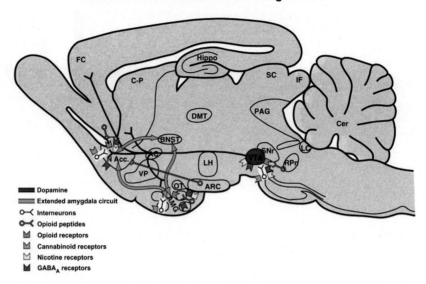

FIGURE 16.1 Sagittal section through a representative rodent brain illustrating the pathways and receptor systems implicated in the acute reinforcing actions of drugs of abuse.

Cocaine and amphetamines activate the release of dopamine in the nucleus accumbens and amygdala via direct actions on dopamine terminals. Opioids activate opioid receptors in the ventral tegmental area, nucleus accumbens, and amygdala via direct actions on interneurons. Opioids facilitate the release of dopamine in the nucleus accumbens via an action either in the ventral tegmental area or the nucleus accumbens (but also are hypothesized to activate elements independent of the dopamine system). Alcohol activates γ-aminobutyric acid-A (GABA$_A$) receptors in the ventral tegmental area, nucleus accumbens, and amygdala via either direct actions at the GABA$_A$ receptor or through indirect release of GABA. Alcohol is hypothesized to facilitate the release of opioid peptides in the ventral tegmental area, nucleus accumbens, and central nucleus of the amygdala. Alcohol facilitates the release of dopamine in the nucleus accumbens via an action either in the ventral tegmental area or the nucleus accumbens. Nicotine activates nicotinic acetylcholine receptors in the ventral tegmental area, nucleus accumbens, and amygdala, either directly or indirectly, via actions on interneurons. Nicotine also may activate opioid peptide release in the nucleus accumbens or amygdala independent of the dopamine system. Cannabinoids activate CB1 receptors in the ventral tegmental area, nucleus accumbens, and amygdala via direct actions on interneurons. Cannabinoids facilitate the release of dopamine in the nucleus accumbens via an action either in the ventral tegmental area or the nucleus accumbens, but also are hypothesized to activate elements independent of the dopamine system. Endogenous cannabinoids may interact with postsynaptic elements in the nucleus accumbens involving dopamine and/or opioid peptide systems. The blue arrows represent the interactions within the extended amygdala

(*Continued*)

more dopamine, thereby causing a prolongation of its pleasurable effects. Cocaine can also block transporters for NE and HT3 as well as binding to and stimulating receptors for dopamine, NE, and HT3. However, its inhibitory effects on the dopaminergic transporter are believed to be its most important mechanism of action.

Cocaine, like many if not all of the drugs to be discussed in this chapter, produces structural changes in brain circuitry. This in turn leads to tolerance, whereby with time the user requires more of the drug to get the same effect. The mechanism for tolerance is caused by an upregulation of the dopamine receptor, producing the requirement for a greater amount of cocaine to produce the same pleasurable feeling. Cocaine also induces dependency, causing the user to require continual access to the drug to avoid numerous unpleasant physical and psychological symptoms categorized as withdrawal.

Cocaine use during pregnancy was initially thought to produce major and devastating effects on the fetus including small brain weight and attention deficit—hyperactivity disorder (ADHD). As researchers began to study the long-term effects of cocaine on the fetus as well as the antenatal ones, the effects were found to be much less profound and most babies totally recover by the time they enter school. However, the cocaine abuser's insatiable desire to obtain the drug on the black market in many cases leads to the neglect of their children. This neglect as well as the cost of obtaining cocaine can have profound negative consequences for the cognitive and behavioral development of the abuser's children.

Amphetamines come from the ephedra plant and were used for medicinal purposes in China over 5000 years ago. After amphetamines were first synthesized in the early 1900s, they were used for a variety of purposes. For example, Benadryl is an amphetamine used as an antihistamine to treat allergies. In the 1960s, amphetamines became popular as weight loss drugs before their potential for abuse was realized. Now amphetamines, including Adderall, are used to treat ADHD.

Like cocaine, amphetamines' major site of action is at the presynaptic terminals of dopaminergic neurons from the ventral segmental area that innervate neurons in the nucleus accumbens. Amphetamines enter the presynaptic

◀ system hypothesized to have a key role in drug reinforcement. *AC*, anterior commissure; *AMG*, amygdala; *ARC*, arcuate nucleus; *BNST*, bed nucleus of the stria terminalis; *Cer*, cerebellum; *C-P*, caudate-putamen; *DMT*, dorsomedial thalamus; *FC*, frontal cortex; *Hippo*, hippocampus; *IF*, inferior colliculus; *LC*, locus coeruleus; *LH*, lateral hypothalamus; *N Acc.*, nucleus accumbens; *OT*, olfactory tract; *PAG*, periaqueductal gray; *RPn*, reticular pontine nucleus; *SC*, superior colliculus; *SNr*, substantia nigra pars reticulata; *VP*, ventral pallidum; *VTA*, ventral tegmental area.

Squire LR, Berg D, Bloom FB, Du Lac S, Ghosh A, Spitzer NC. Fundamental Neuroscience. 4th ed. Oxford, UK: Elsevier; 2013 [Chapter 41]. Fig. 41.8, p. 488. With permission.

endings of these cells through the dopamine transporters, thereby causing dopamine release into the synapse. While amphetamines can produce tolerance and dependency in heavy users, there is no evidence that damage to the fetus occurs from amphetamine use. As is true of cocaine, the secondary consequences of amphetamine abuse on childhood cognitive and social development can be very serious.

Opioids are another important class of drugs similar to cocaine in that they work in part through stimulation of the pleasure center, but they have other actions as well. Initially opium was produced from the pods of the opium poppy, and has been used as a treatment for pain as well as for recreational purposes for many years. Morphine and diamorphine (heroin) were isolated from poppy pods in the 1800s and were used soon thereafter to relieve acute or chronic severe pain. Other semisynthetic or synthetic morphine-like substances including oxycodone, fentanyl, naloxone, and methadone were synthesized in the 20th century.

The last 50 years have witnessed an explosion of knowledge concerning the mechanism of action of opioids and the discovery of endogenous opioids and their receptors. There are three major classes of opioid receptor—μ, δ, and — which are all members of the GPCR superfamily. Members of the endogenous opioid family members composed of enkephalins, beta endorphin, and dynorphin have a number of important physiological effects, including alleviation of pain. The rest of this discussion, however, will mainly concern μ receptors, which are found throughout the pleasure or reward circuits on γ-aminobutyric acid (GABA) ergic interneurons. Some of the μ receptors, when bound to their ligand, inhibit the dopaminergic neurons that project to the nucleus accumbens. When an opioid binds to the μ receptors on these neurons, they inhibit their neurotransmitter release, which results in the increased firing of the rewarding, pleasurable dopaminergic neurons. Like other drugs of potential abuse, continued usage leads to tolerance and dependency. Withdrawal from opioids produces a number of unpleasant and potentially lethal effects. Also, overdoses of the more potent opioids like fentanyl, or combining opioid and depressant drug use (e.g., heroin and alcohol) can produce respiratory failure and death as has occurred in the recent opioid epidemic in the United States.

There have not been many research studies into the consequences of both legal and illegal opioid usage during pregnancy, on both prenatal and antenatal brain development, but this has begun to change. There is evidence in animals that opioid usage during pregnancy can cause changes in the timing of myelination and dendrite morphogenesis. In humans, regional brain volume decreases in the basal ganglia and other brain areas have been reported.

Another major problem with infants born to mothers exposed to opioids during pregnancy is that they are born dependent on opioids and immediately suffer from neonatal abstinence syndrome. These babies must remain hospitalized until withdrawal is complete and they are more likely to require further hospitalizations. Also there appear to be long-term decreases in IQ and other cognitive measures in older children.

Finally, opioid-addicted parents of children are notoriously bad parents, subjecting their children to abuse and/or neglect. Both parental behaviors have very negative consequences for the child's intellectual and behavioral development.

Two legal drugs that are also known to produce damage to the fetal brain if used during pregnancy are alcohol and tobacco. Both these drugs indirectly affect the so-called pleasure center in the nucleus accumbens, but they also have other receptor-mediated sites of action.

Alcohol is produced by the fermentation of a variety of fruits and vegetables. Humans have been producing fermented alcoholic beverages for many thousands of years. An example of the popularity of alcohol-containing drugs is the almost universal drinking of apple cider until the late 19th century. This was in part due to the contamination of the water supply. Drinking of apple cider and other alcoholic beverages caused many individuals to become "drunkards." This in turn led to the formation of various temperance societies. The temperance societies were very successful in preaching the evils of alcohol, and their increasing influence eventually led to passage of the 18th Amendment to the US Constitution. This amendment made the selling of alcohol illegal. Because of the development of a strong black market and the increase in crime created by this amendment, it was repealed by the 21st Amendment.

While alcohol causes a short-term pleasurable feeling, it has depressant activities leading to drowsiness and decreased executive function. Chronic alcohol use almost invariably leads to tolerance and dependency together with difficulty withdrawing. It has also proven very difficult for dependent individuals to abstain. Among the many reasons for this is the pervasive presence of legal alcohol in our society.

Chronic alcoholics suffer from liver and pancreatic damage. Also, due to their eating a reduced diet with too little vitamin B12, many suffer from a specific brain lesion called Korsakov's syndrome. This syndrome is characterized by a memory deficit combined with flagrant confabulation. There is also thought to be a higher risk of dementia from heavy drinking, although moderate consumption of wine may protect from dementia.

For many years it was believed that alcohol exerted its effects on the brain through a generalized perturbing action due to its relatively hydrophobicity compared with that of water. In recent years, however, scientists have identified a site on the ionic $GABA_A$ receptor, which activates this receptor and is thought to mediate many of alcohol's effects on the brain. As we know, this receptor is the major inhibitory neurotransmitter in the brain, and the depressant effects of alcohol can be explained by this fact. Also, alcohol is a specific inhibitor of the N-methyl-d-aspartate (NMDA) receptor, which is very important in memory. It is also likely that alcohol indirectly affects the dopaminergic pleasure center in the nucleus accumbens by activating the $GABA_A$ receptor.

As we have discussed in previous chapters, both GABA and Glu function to induce synapses in the mammalian brain during development. Inhibition of these receptors, which play important roles in fetal brain development, may produce alcohol spectrum disorder.

Alcohol spectrum disorder is caused by a pregnant woman drinking two or more drinks a day during pregnancy. The most extreme version, produced by consumption of four or more drinks a day, is fetal alcohol syndrome (FAS). Approximately between 2% and 10% of women in the United States drink alcohol during pregnancy. The lifetime cost of each individual with FAS is estimated to be $2,000,000.

FAS produces devastating effects on the developing brain that persist throughout life. During the first trimester of pregnancy, alcohol interferes with neuronal migration and organization and can produce microcephaly. In the third trimester, alcohol damages the hippocampus, which produces memory deficits and epilepsy. These fetal insults lead in many cases to poor memory, decreased executive functioning, learning disabilities, and poor social skills among other deficits. There is no treatment for FAS. It can be entirely prevented, however, by the complete avoidance of alcohol throughout the entire course of pregnancy. Again, as with the other drugs of abuse, an alcoholic parent(s) can often abuse or neglect their children, causing a wide variety of cognitive and behavioral consequences.

Tobacco is the second legal drug that has major consequences to the fetal brain if used during pregnancy. Tobacco has been employed as a recreational drug by Native Americans for thousands of years. It is a stimulant, producing heightened alertness and a general feeling of well-being as well as decreased anxiety. Unfortunately, it produces both tolerance and dependency. Withdrawal is extremely difficult and leads to many other problems including depression and weight gain.

Nicotine is the primary active ingredient in tobacco. It binds selectively to the nicotinic ACh receptor, which is found in many areas of the brain and the spinal cord. Nicotine functions as an agonist causing opening of a channel that allows Na^+ to enter the cytoplasm and K^+ to exit. This causes neurons to be depolarized. A primary effect is to increase the release of dopamine from ventral segmental neurons in the nucleus accumbens, thereby stimulating the pleasure center of the brain.

There is evidence that tobacco use during pregnancy leads to structural changes in the brain and to smaller brain weight at birth. There are also reports of numerous cognitive and behavioral changes in older children resulting in problems in learning and auditory difficulties.

As described in this chapter, all of the addictive drugs share similar mechanisms. Fig. 16.2 summarizes the molecular mechanisms involved with all the addictive drugs.

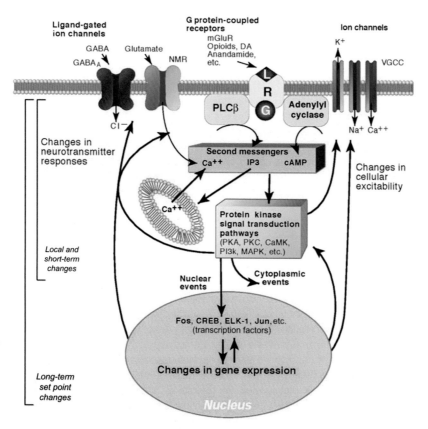

FIGURE 16.2 Molecular mechanisms of neuroadaptation.

Cocaine and amphetamines, as indirect sympathomimetics, stimulate the release of dopamine, which acts at G protein-coupled receptors, specifically D1, D2, D3, D4, and D5. These receptors modulate the levels of second messengers like cyclic adenosine monophosphate (cAMP) and Ca^{2+}, which in turn regulate the activity of protein kinase transducers. Such protein kinases affect the functions of proteins located in the cytoplasm, plasma membrane, and nucleus. Among membrane proteins affected are ligand-gated and voltage-gated ion channels (VGCCs). Gi and Go proteins also can regulate K^+ and Ca^{2+} channels directly through their $\beta\gamma$ subunits. Protein kinase transduction pathways also affect the activities of transcription factors. Some of these factors, like cAMP response element binding protein (CREB), are regulated posttranslationally by phosphorylation; others, like Fos, are regulated transcriptionally; still others, like Jun, are regulated both posttranslationally and/or transcriptionally. While membrane and cytoplasmic changes may be only local (e.g., dendritic domains or synaptic boutons), changes in the activity of transcription factors may result in long-term functional changes. These may include changes in gene expression of proteins involved in signal transduction and/or neurotransmission, resulting in altered neuronal responses. For example, chronic exposure to psychostimulants has been reported to increase levels of protein kinase A (PKA) and adenyl cyclase in the nucleus and to decrease levels of Gαi. Chronic exposure to psychostimulants also alters the expression of transcription factors themselves. CREB expression, for instance, is depressed in the nucleus accumbens by chronic cocaine

(Continued)

◀ treatment. Chronic cocaine induces a transition from Fos induction to the induction of the much longer-lasting Fos-related antigens such as ΔFosB. Opioids, by acting on neurotransmitter systems, affect the phenotypic and functional properties of neurons through the general mechanisms outlined in the diagram. Shown are examples of ligand-gated ion channels such as the γ-aminobutyric acid-A ($GABA_A$) and glutamate NMDA receptor (NMR) and G protein-coupled receptors such as opioid, dopamine (DA), or cannabinoid CB1 receptors, among others. These receptors modulate the levels of second messengers like cAMP and Ca^{2+}, which in turn regulate the activity of protein kinase transducers. Chronic exposure to opioids has been reported to increase levels of PKA and adenyl cyclase in the nucleus accumbens and to decrease levels of $G\alpha i$. Chronic exposure to opioids also alters the expression of transcription factors themselves. CREB expression, for instance, is depressed in the nucleus accumbens and increased in the locus coeruleus by chronic morphine treatment, while chronic opioid exposure activates Fos-related antigens such as ΔFosB. Alcohol, by acting on neurotransmitter systems, affects the phenotypic and functional properties of neurons through the general mechanisms outlined in the diagram. Shown are examples of ligand-gated ion channels such as the $GABA_A$ and the NMDA receptor and G protein-coupled receptors such as opioid, dopamine, or cannabinoid CB1 receptors, among others. The latter also are activated by endogenous cannabinoids such as anandamide. These receptors modulate the levels of second messengers such as cAMP and Ca^{2+}, which in turn regulate the activity of protein kinase transducers. Such protein kinases affect the functions of proteins located in the cytoplasm, plasma membrane, and nucleus. Among membrane proteins affected are ligand-gated and VGCCs. Alcohol, for instance, has been proposed to affect the $GABA_A$ response via protein kinase C (PKC) phosphorylation. Gi and Go proteins also can regulate K^+ and Ca^{2+} channels directly through their $\beta\gamma$ subunits. Chronic exposure to alcohol has been reported to increase levels of PKA and adenyl cyclase in the nucleus accumbens and to decrease levels of $Gi\alpha$. Moreover, chronic ethanol induces differential changes in subunit composition in the $GABA_A$ and glutamate inotropic receptors and increases expression of VGCCs. Chronic exposure to alcohol also alters the expression of transcription factors themselves. CREB expression, for instance, is increased in the nucleus accumbens and decreased in the amygdala by chronic alcohol treatment. Chronic alcohol induces a transition from Fos induction to the induction of the longer-lasting Fos-related antigens. Nicotine acts directly on ligand-gated ion channels. These receptors modulate the levels of Ca^{2+}, which in turn regulate the activity of protein kinase transducers. Chronic exposure to nicotine has been reported to increase levels of PKA in the nucleus accumbens. Chronic exposure to nicotine also alters the expression of transcription factors themselves. CREB expression, for instance, is depressed in the amygdala and prefrontal cortex and increased in the nucleus accumbens and ventral tegmental area. Δ9-Tetrahydrocannabinol (THC), by acting on neurotransmitter systems, affects the phenotypic and functional properties of neurons through the general mechanisms outlined in the diagram. Cannabinoids act on the cannabinoid CB1 G protein-coupled receptor. The CB1 receptor also is activated by endogenous cannabinoids such as anandamide. This receptor modulates (inhibits) the levels of second messengers like cAMP and Ca^{2+}, which in turn regulate the activity of protein kinase transducers. Chronic exposure to THC also alters the expression of transcription factors themselves. *CaMK*, Ca^{2+}/calmodulin-dependent protein kinase; *ELK-1*, E-26-like protein 1; *PLCβ*, phospholipase C β; *IP3*, inositol triphosphate; *MAPK*, mitogen-activated protein kinase; *PI3K*, phosphoinositide 3-kinase; *R*, receptor.

Squire LR, Berg D, Bloom FB, Du Lac S, Ghosh A, Spitzer NC. Fundamental Neuroscience. 4th ed. Oxford, UK: Elsevier; 2013 [Chapter 41]. Fig. 41.12, p. 894. With permission.

RECOMMENDED BOOK CHAPTERS

Kandel ER, Schwartz JH, Jessel TM, Siegelbaum SA, Hudspeth AJ. *Principles of Neural Science*. 5th ed. New York: McGraw Hill; 2013 [Chapter 49].

Squire LR, Berg D, Bloom FB, Du Lac S, Ghosh A, Spitzer NC. *Fundamental Neuroscience*. 4th ed. Oxford, UK: Elsevier; 2013 [Chapter 41].

Review Articles

Benowitz NL. Pharmacology of nicotine: addiction, smoking-induced disease, and therapeu-tics. *Annu Rev Pharmacol Toxicol*. 2009;49:57−71.

Carvalho M, Carmo H, Costa VM, et al. Toxicity of amphetamines: an update. *Arch Toxicol*. 2012;86:1167−1231.

Hummel M, Unterwald EM. D1 dopamine receptor: a putative neurochemical and behavioral link to cocaine action. *J Cell Physiol*. 2002;191:17−27.

McIntosh C, Chick J. Alcohol and the nervous system. *J Neurol Neurosurg Psychiatry*. 2004;75(Suppl III):16−21.

Wilson-Poe AR, Jeong HJ, Vaughan CW. Chronic morphine reduces the readily releasable pool of GABA, a presynaptic mechanism of opioid tolerance. *J Physiol*. 2017;595:6541−6555.

Role of trophic factors and receptors in developmental disorders

17

CHAPTER OUTLINE

A number of developmental brain disorders involve known genetic causes and many of them involve receptors or trophic factors. A group of disorders that affect lysosomes, the cell's digestive organelle, are caused by defects in enzymes, located in lysosomes, which digest a plethora of specific mucopolysaccharides and lipids. One rare lysosomal enzyme disorder, I-cell disease, produces many developmental problems in the brain and other organs. This disease is caused by a deficiency in the enzyme N-acetyl glucosamine transferase, which transfers the carbohydrate Mannose-6-phosphate to soluble lysosomal enzyme precursors in the *cis*-Golgi. When the lysosomal enzymes containing the Mannose-6-phosphate tag reach the *trans*-Golgi network, they bind to the Mannose-6-phosphate receptor. After the lysosomal enzymes and the Mannose-6-phosphate receptors are enclosed in clathrin-coated vesicles, they are transported to the late endosome/lysosome (Fig. 17.1).

In individuals with I-cell disease the lysosomal enzyme precursors do not receive the Mannose-6-phosphate tags. Therefore, they are secreted from the cells instead of being transported to the lysosome. I-cell disease leads to a severe impairment of brain development and resultant intellectual abilities.

Fragile X syndrome is an X-linked disease that leads to intellectual and social impairment. It is found in 1 in 4000 male births and in 1 in 8000 female births. The condition is much worse in males, who have only one X chromosome and therefore have no functional fragile X mental retardation 1 (FMR1) protein. Females have two X chromosomes and produce 50% of the normal amount of the functional protein.

The mechanism by which the disease is produced involves the CGG triplet found in the FMR1 5′ nontranslated portion of the DNA. Normal individuals have 5−44 CGG repeats in the DNA that code for the FMR1 protein. However, when the CGG repeats expand to over 200, the resulting protein can no longer perform its function and the disease phenotype is expressed.

FMR1 is mainly produced in neurons in the brain. Its function is to bind to a select group of mRNAs, approximately 4% of the total. This binding takes place in the nucleus. Hence FMR1 can be considered as a receptor that binds a particular structure on these mRNAs. FMR1 then transports the mRNAs to dendritic

Receptors in the Evolution and Development of the Brain. DOI: https://doi.org/10.1016/B978-0-12-811012-6.00017-0

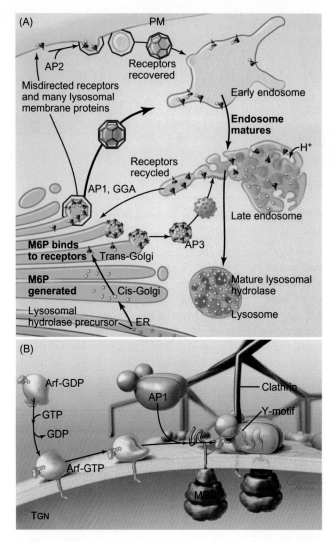

FIGURE 17.1 Sorting pathways used by mannose-6-phosphate receptors and coat assembly at the *trans*-Golgi network.

(A) Mannose-6-phosphate receptors (MPRs) carry newly synthesized lysosomal hydrolases with mannose-6-phosphate (M6P) from the *trans*-Golgi network (TGN), via endosomes, to lysosomes, after which the MPRs return to the TGN. Receptors misdirected to the cell surface are recovered by endocytosis and returned to the pathway in endosomes. GGA (Golgi-localizing, γ-adaptin ear domain homology, Arf-binding protein), clathrin adapter proteins. (B) Coordination of coat assembly and cargo recruitment at the TGN. An exchange factor activates the small GTPase Arf to bind GTP, which triggers recruitment of adaptor protein 1 (AP1) coat constituents to the TGN membrane. The MPR is concentrated in the emerging coated vesicle through interactions between a tyrosine-based sorting motif in its cytoplasmic domain and the μ-subunit of AP1.

Pollard TD, Earnshaw WC, Lippincott-Schwartz J, Johnson GT. Cell Biology. 3rd ed. Philadelphia, PA: Elsevier; 2017 [Chapters 1, 3, 13–16, 21–24, 27, 30, 33, and 34]. Fig. 21.25, p. 373. With permission.

terminals where they can be translated into proteins. Among the proteins whose synthesis in this is decreased in the fragile X syndrome is the metabotropic glutamate receptor. FMR1 also directly binds to Na + and Ca + + activated K + channels. These proteins are important for long-term potentiation (LTP) and long-term depression (LTD) and the resultant memory acquisition and learning (Fig. 17.2). Dopaminergic receptors, which are involved in developing social

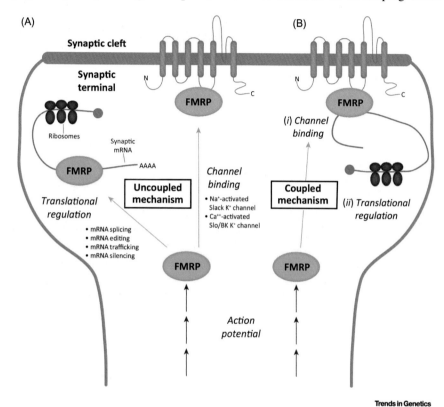

Trends in Genetics

FIGURE 17.2 Roles of fragile X mental retardation protein (FMRP) in RNA and channel binding at the neuronal synapse.

Activity-dependent functions of FMRP in RNA-binding translation regulation and direct channel-binding activity regulation. In the uncoupled mechanism (A), RNA- and channel-binding roles are unrelated, representing two evolutionarily divergent functions. In the coupled mechanism (B), channel binding (i) is an integral activity-sensing step in the translational regulation of FMRP-bound transcripts (ii). Activity-dependent functions of FMRP in RNA-binding translation regulation and direct channel-binding activity regulation. In the uncoupled mechanism (A), RNA- and channel-binding roles are unrelated, representing two evolutionarily divergent functions. In the coupled mechanism (B), channel binding (i) is an integral activity-sensing step in the translational regulation of FMRP-bound transcripts (ii).

Davis JK, Broadie K. Multifarious functions of the fragile X mental retardation protein. Trends Genet. *2017;33:703 – 714. PMID: 28826631. Fig. 1. With permission.*

skills, are also affected. This may account for the high rate of attention disorders and hyperactivity in children with the disease. Finally, $GABA_A$ receptor downregulation may account for anxiety symptoms manifested in children with fragile X syndrome.

Down syndrome is a common genetic disease found in 1 in 1000 live births. Its frequency rises from 0.3% in mothers 20 years old to 3% in mothers over 45. It is caused by an extra chromosome 21. The triplication occurs when the chromosomes don't separate properly during egg or sperm development. However, the condition is much more common during the development of the egg. Chromosome 21 contains 310 genes, all of which are triplicated in the disease. Different regions of the chromosome triplication produce separate phenotypes including craniofacial abnormalities, heart abnormalities, digestive difficulties, and cognitive defects.

The triplication of a number of receptors found on chromosome 21 is implicated in the cognitive and behavioral deficits seen during development and in later years in patients with Down syndrome. Perhaps the most interesting triplication is that of the beta amyloid precursor protein (APP). This protein will be discussed in detail in the next chapter with respect to its role in the etiology of Alzheimer's disease (AD).

APP is a one-pass transmembrane protein with most of its mass on the extracellular side of the plasma membrane. It moves through the cell in vesicles with its small cytoplasmic domain functioning as a receptor linking transport vesicles to kinesins, the motor protein family, members of which can move vesicles in either a forward or backward direction (See Chapter 5: Neuronal cell biology). There is now a mouse model of Down syndrome that produces 50% more APP than wild-type. In these mice the retrograde transport of nerve growth factor (NGF), the trophic factor necessary for the survival of basal forebrain cholinergic neurons, is compromised (Fig. 17.3). This leads to the death of a portion of these cells. It is known that the death of cholinergic basal forebrain neurons also occurs in the early stages of AD.

As the care of patients with Down syndrome has become more sophisticated, these individuals are living into their fifties. Unfortunately they begin to show the symptoms of AD in their forties. AD will be discussed in the next chapter.

Another receptor that could play an important role in Down syndrome is CAM2, a member of the immunoglobulin (Ig) superfamily, which contains five Ig domains and is anchored to the cell membrane through either a transmembrane sequence or a glypiated (GPI) residue. CAM2 mainly functions as a homophilic receptor, recognizing an identical molecule on the surface of another neuron. This causes the bundling of neurites, either axons or dendrites, during development. Excess CAM2 in Down patients can produce developmental deficits. Several other receptors also may play roles in the cognitive problems seen in Down syndrome. The recent advent of stem cells from Down patients that can be differentiated into neurons should allow more rapid progress in our understanding of the role of receptors and other proteins in this disease. Human stem cells and their value in unraveling the cause of many human diseases are discussed in

(A) Structure of a nerve cell

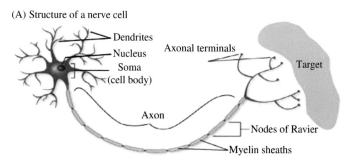

(B) Different types of NGF signaling endosomes

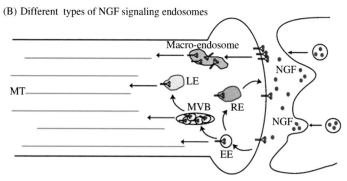

FIGURE 17.3

(A) A schematic representation of a nerve cell. A typical neuron possesses an extensive network of dendrites and axons. The axon is often insulated with the myelin sheaths that are interrupted by the nodes of Ranvier. The axon terminal innervates the target. The neurotrophic signal, released from the target, must be retrogradely transported to the cell body (soma). (B) The different types of signaling endosomes. Neurotrophins, such as nerve growth factor (NGF), are released from the target and bind to the Trk receptors at the axonal terminals. The ligand and receptor complex is internalized and is shuttled via the endocytic pathways. Endocytic compartments such as the early endosomes, MVBs, and late endosomes have all been proposed to traffic the neurotrophic signal to the cell body. Through a different internalization pathway, macroendosomes may also be formed to carry the neurotrophin signals retrogradely.

Wu C, Cui B, He L, Chen L, Mobley WC. *The coming of age of axonal neurotrophin signaling endosomes.* J Proteomics. *2009;72:46−55. PMID: 19028611. Fig. 1. With permission.*

Chapter 19, Potential use of neuronal stem cells to replace dying neurons depends on trophic factors and receptors.

Until recently, diseases including schizophrenia, bipolar disease and depression, attention hyperactivity disorder, and autism spectrum disorder (ASD), which usually manifest during the first two decades of life, were considered to be psychiatric disorders with no organic brain damage. However, more recently, careful autopsies together with genetic, behavioral, and animal model studies have found underlying neuropathology in all four disorders. In each case, trophic factor and receptor changes have been associated with the disease.

Bipolar disease has about an 85% risk of being familial, even though no gene or genes have been conclusively identified as causing the disease. Two drugs, lithium and valproic acid, have been used for many years for treatment with very good effect. These drugs show potent biochemical effects both on isolated neurons and in vivo, including potent neurotrophic and neuroprotective effects. Valproic acid was first used to treat epilepsy and then discovered to be an excellent mood stabilizer at slightly lower doses.

Both lithium and valproic acid activate pathways including the MAP kinase and ERK pathways, which are known to be stimulated by neurotrophins including BDNF. The Wnt family of proteins, which play key roles during neural development, also signal through the same pathways as do lithium and valproic acid.

Besides stimulating the same pathways as do neurotrophins and Wnts, lithium and valproic acid also increase the synthesis of a number of neurotrophic factors including BDNF, glial-derived growth factor, and vascular endothelial growth factor. They also increase the synthesis of antiapoptotic proteins including members of the Bcl-2 family.

In models of both bipolar disease and depression in rodents and primates, antidepressants including fluoxitine and riboxetine produce neurogenesis in hippocampal neurons together with increased hippocampal dependent learning and associated LTP. These effects also appear to be mediated through signaling pathways involving MAPK, ERK, and Wnts.

There is a fairly common variant of the gene for BDNF that substitutes a Met for Val in the 66th amino acid. Individuals with one copy of the Val66Met allele don't secrete as much BDNF as do normal individuals. Treatment with antidepressants that alleviate the symptoms of both bipolar disease and depression produces more vigorous BDNF secretion in patients with either disease.

Schizophrenia, which usually begins in adolescence or early adulthood, also has a strong hereditary component. Neuropathological examination of the brains of schizophrenics who died soon after being diagnosed show a significant decrease in synapses in areas of the brain associated with the disease. Presumably there is an increased level of synaptic pruning in schizophrenics compared with that in normals.

Recently, a possible explanation for the increase in synaptic pruning has been discovered. A large genetic screen of people with schizophrenia picked out a gene, C4, which is found in the MHC region of the human genome. This region contains many genes involved in immune surveillance, including C4. The protein product of this gene, C4, is found on many synapses in parts of the developing brain that are affected in schizophrenia. There is evidence that microglia use this protein as a receptor and destroy the synapses that contain it. People with too much synaptic C4 are at high risk for developing schizophrenia due to excess destruction of the C4-containing synapses.

Antipsychotic drugs are now used to treat schizophrenia. Classical antipsychotics, including haldol and reserpine, mainly act at dopaminergic receptors. The more recently described atypical antipsychotics bind to a much broader range of receptors. These include those for dopamine, serotonin, and glutamate. There is both in vitro and in vivo evidence that these two classes of antipsychotics show neuroprotective and neurotrophic effects that may play a role in the relief of various schizophrenic behaviors. A number of signaling cascades are stimulated by these agents.

ASD is defined as a group of related disorders characterized by poor social communication and the presence of restrictive, persistent modes of behavior. The incidence of this disorder has increased dramatically in the last 30 years. ASD is highly heritable. Siblings of children with ASD have a 100-fold greater chance of having ASD than do siblings of normals.

Molecular and genetic techniques have enabled the identification of a number of gene variants associated with ASD. There is a strong association of ASD with two genes coding for receptors that are important during brain development, the Neurexins and the metabotropic GABAβ3 receptor.

Two additional genes that have been associated with ASD are those that code for oxytocin (OX) and arginine vasopressin (AVP). These two hormones are produced by different groups of cells in the hypothalamus. Mutations in both genes have been implicated as risk genes for ASD. These two hormones and their receptors have been discussed in Chapter 13, Steroid hormones and their receptors are key to sexual differentiation of the brain while glucocorticoids and their receptors are key components of brain development and stress tolerance; the hypothalamic hormone producing cells: importance in various functions during and after brain development. OX, AVP, and their respective receptors have been associated with a variety of social behaviors in animal studies and human studies. A large genetic study has implicated OX receptor variants in ASD and generalized social impairment. Plasma OX has been found to be lower in children with ASD. Also OX and serotonin levels were found to be negatively correlated with each other in children with ASD.

RECOMMENDED BOOK CHAPTERS

Kandel ER, Schwartz JH, Jessel TM, Siegelbaum SA, Hudspeth AJ. *Principles of Neural Science*. 5th ed. New York: McGraw Hill; 2013 [Chapters 62, 63, 64].

Squire LR, Berg D, Bloom FB, Du Lac S, Ghosh A, Spitzer NC. *Fundamental Neuroscience*. 4th ed. Oxford, UK: Elsevier; 2013 [Chapters 43, 50].

Review Articles

Hagerman RJ, Berry-Kravis E, Hazlett HC, et al. Fragile X syndrome. *Nat Rev Dis Primers*. 2017;3:17065.

Hong CJ, Liou YJ, Tsai SJ. Effects of BDNF polymorphisms on brain function and behavior in health and disease. *Brain Res Bull*. 2011;86:287–297.

Kasem E, Kurihara T, Tabuchi K. Neurexins and neuropsychiatric disorders. *Neurosci Res.*. 2018;127:53–60.

Sekar A, Bialas AR, de Rivera H, et al. Schizophrenia risk from complex variation of complement component 4. *Nature*. 2016;530:177–183.

Neuronal survival and connectional neurodegenerative diseases

18

CHAPTER OUTLINE

One of the great triumphs of modern medicine is the prevention and treatment of infectious diseases such as smallpox, tuberculosis, and polio. Unfortunately, however, the resulting aging of the population has led to a great increase in chronic diseases including heart disease, cancer, type II diabetes, and a number of brain diseases.

The neurons of the brain, with the exception of the nasal epithelial cells and the cells of the olfactory bulb that innervate them, and the neurons comprising the dentate gyrus of the hippocampus, are required to survive for the length of an individual's life. While synaptic plasticity allows compensation for some neuronal loss, the destruction of a relatively small number of neurons in key regions can produce various impairments and even death.

In 1981 Stanley Appel coined the phrase "connectional diseases of aging" to describe a class of neuronal diseases that start in one small locus and slowly progress by spreading sequentially to groups of neurons in other regions. (See General References.) The postulate discussed in Chapter 7, A testable theory to explain how the wiring of the brain occurs—that is, that some trophic support is necessary throughout life for neurons, and that this support comes via synaptic release of trophic factors during neuronal firing—provides a fundamental understanding of a number of age-related, connectional, neurodegenerative diseases. These include Alzheimer's (AD), Parkinson's (PD), amyotrophic lateral sclerosis (ALS), prion diseases, and Huntington's disease (HD). These are all characterized by the relatively slow spreading from the initial nidus of the disease to multiple regions that are synaptically connected, and therefore trophically connected. In contrast, neurons that are adjacent to dying neurons but not synaptically connected are spared any damage.

Many of the diseases discussed in this chapter have their origins in a specific region. For example, AD is thought to begin in the locus ceruleus, while PD is believed to begin in the medulla and olfactory bulb. The most obvious possibility is that the mutant or difunctional protein is more highly expressed in the region of the brain in which the disease originates. However, in many cases the suspected causal protein is not found in a greater concentration in the brain area in which the disease originates than in other regions. Therefore if we hope to treat these diseases with stem cell replacement therapy, which will be discussed in the next chapter, or with other treatments, the origin of each disease will provide us with crucial data. This information in turn may lead to the discovery of the optimum period in which to begin treatment, to minimize the synaptic and neuronal loss.

An analogous situation is confronted in the case of heart disease. The cardiac muscle cells, which comprise the majority of the heart cells, do not divide in the adult. It was determined that high cholesterol and high blood pressure were great risk factors for heart disease. Therefore, drugs such as statins, which lower blood cholesterol, as well as a variety of drugs that lower blood pressure were introduced to prevent damage to the heart before it occurs. These drug therapies combined with lifestyle changes have resulted in a marked decrease in deaths from heart disease. One would hope that a therapy that can be applied before symptoms appear will also be important in the prevention of connectional diseases.

I will now discuss several of these age-related, neurodegenerative diseases, and the roles of trophic factors and/or receptors in their etiology and/or spread. AD is the most common age-related neurodegenerative disease. It affects approximately one-eighth of all individuals over age 65. The number of people with the disease is estimated to triple in the next 25 years because the major risk factor for AD is aging, and the population of aged people will also likely triple during this time span. Patients in the final phase of this disease, at present an invariably fatal illness, will require 24-hour care either at home or in a nursing facility. Therefore AD will become an enormous economic burden on society if a prevention and/or treatment is not rapidly discovered.

AD was first described in 1901 in a 41-year-old woman. After death an autopsy was performed on her brain, which revealed three characteristic pathologies. On gross inspection the brain was shrunken with much larger ventricles. Another pathology seen using histological methods on thin sections of affected regions was the deposition of extracellular plaques that are made up predominantly of a protein, amyloid beta (Aβ) (Fig. 18.1). This peptide is cleaved from a larger precursor molecule called the beta amyloid precursor protein (APP). These proteins were discussed in the previous chapter with respect to Down syndrome.

Aβ plaques form in almost all cortical regions of the brain. These extracellular plaques form a very insoluble component called amyloid, which is composed of cross β pleated sheets. (See General References.) Amyloid fibers are found in many parts of the body as well as in the brain and can be composed of many

(A)   (B)

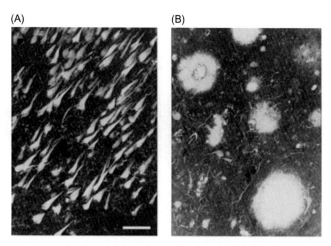

FIGURE 18.1 Photomicrographs of neurofibrillary tangles (A) and amyloid-containing plaques (B) in the hippocampal formation of a patient who died with late-stage Alzheimer's disease.

Note that the size and morphological characteristics of plaques in (B) vary widely. Scale bar = 50 μm.

Squire LR, Berg D, Bloom FB, Du Lac S, Ghosh A, Spitzer NC. Fundamental Neuroscience. 4th ed. Oxford, UK: Elsevier; 2013; [Chapters 28, 30 43, 48]. Fig. 43.6, p. 941. With permission.

different proteins. The Aβ plaques appear to be the first indication of AD pathology, although many elderly people whose brains contain these plaques do not have clinical symptoms of AD.

The third lesion found in the brains of AD patients is the intracellular accumulation of neurofibrillary tangles (NFTs), composed of the tau protein (Fig. 18.1). Tau is normally found only in axons, where it serves to crosslink microtubules, but it appears that early in the course of the disease tau falls off the microtubules and forms NFTs. These consist of hyperphosphorylated twisted fibers, and are present in all three cellular compartments (Fig. 18.2). NFTs are also composed of amyloid fibers.

NFTs first appear in particular projection neurons in the locus ceruleus. This nucleus contains norepinephrine neurons whose axons project to many cortical regions. NFTs then progress from the neurons of the locus ceruleus sequentially to specific neurons in the entorhinal cortex. They are then seen in neurons in particular regions of the hippocampus. NFT location appears to coincide precisely with the progression of the disease, which is not true of the Aβ plaques. In all these regions in which NFTs sequentially appear, neurons not synaptically/trophically connected to the affected neurons are spared. The time period during which plaques and tangles form with no concomitant clinical symptoms appears to be at least 10 years.

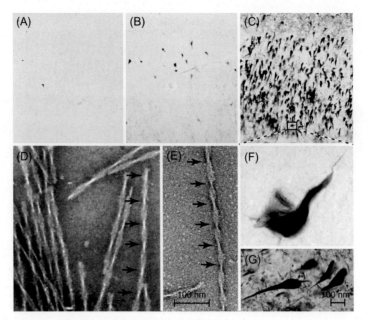

FIGURE 18.2 Neurofibrillary tangles of paired helical filaments in the brains of patients with Alzheimer's disease.

(A—C) Light micrographs of sections of the hippocampus of human brains stained with silver for neurofibrillary tangles. (A) Stage I with few tangles. (B) Stage III with moderate numbers of tangles. (C) Stage V, advanced Alzheimer's disease, showing abundant tangles. (D) Electron micrograph of paired helical laments isolated from Alzheimer neurofibrillary tangles and prepared by negative staining. (E) Electron micrograph of a negatively stained, paired, helical lament reassembled in vitro from recombinant tau protein. (F and G) High-powered micrographs of neurofibrillary tangles from the brain of an Alzheimer patient stained brown with an antibody to tau.

Pollard TD, Earnshaw WC, Lippincott-Schwartz J, Johnson GT. Cell Biology. 3rd ed. Philadelphia, PA: Elsevier; 2017 [Chapters 1, 3, 13—16, 21—24, 27, 30, 33, and 34]. Fig. 34.12, p. 602. With permission.

By the time the NFTs become abundant in the hippocampus, the clinical symptoms, mainly slight memory deficits, begin to appear. This early form of memory loss is called minimal cognitive impairment (MCI). Approximately 50% of individuals with MCI will progress to AD. During progression to AD, overt inflammation occurs including activation of resident microglia and release of proinflammatory proteins. The disease then inexorably progresses to other cortical and subcortical regions, leading inevitably to death.

The key roles of trophic factors and receptors in maintaining the functions and synaptic connectivity and eventually in keeping the neurons alive throughout their life fits in well with the evidence described previously that the first pathology occurs in the locus ceruleus. This damage is followed by destruction of synaptically connected neurons in the entorhinal cortex and hippocampus, and finally to tropically connected neurons in other cortical areas. AD can therefore considered a connectional disease.

The most widely accepted theory for the primary cause of AD is the amyloid hypothesis. This theory posits that the protein that forms the amyloid deposits, Aβ, is the agent that triggers the pathology leading to first to MCI and subsequently to AD. There is much evidence to support this hypothesis although it remains controversial.

The precursor to Aβ, APP, is a plasma membrane integral protein. It functions as a receptor for a number of proteins involved in rapid axonal transport as well as in other critical processes. When cleaved at the α site, which is in the middle of the Aβ sequence on the plasma membrane of the APP molecule, it can function as a trophic factor. The α cleavage is thought to be the normal cleavage site. APP can also be cleaved at another site, the β site, which is proximal to the N-terminal amino acid of Aβ, by an enzyme called β-secretase. It is subsequently cleaved by another enzyme, the γ secretase, which has its active site within the plasma membrane bilayer. This cleavage generates Aβ as well as cytoplasmic fragments that can migrate to the nucleus to turn on transcription of various genes. (See General References.) These cleavages take place in intracellular compartments, predominantly in endosomes.

The 40−amino acid form of Aβ is secreted in normal individuals. This molecule does not appear to be amyloidogenic or toxic to cells. In fact, it may serve an important function in neurons. Aβ 1-40 is present in peptide-containing vesicles at synapses and secreted from pre- and postsynaptic vesicles. This in turn results in the inhibition of further neurotransmitter release and can inhibit LTP. (See General References.) Therefore Aβ 1-40 can function as the yin to the yang of BDNF, which produces increased neurotransmitter secretion and increases LTP. Aβ1-40 can also serve to protect the brain against a variety of microbial infections. These small Aβ species first bind to microbial cell wall proteoglycans. Then, the developing protofilaments inhibit pathogen binding to the host neurons (See General References).

Another form of Aβ, Aβ 1-42 also can be generated by the γ-secretase. This form is normally present in low concentrations compared with Aβ 1-40. However, mutations in the γ-secretase on either of the two homologous chromosomes coding for the protein increase the ratio of Aβ 1-42 to Aβ 1-40. This in turn produces an autosomal dominant form of AD. Therefore, children of the affected individuals have a 50% chance of contracting the disease. Also, a number of mutations either within or adjacent to the Aβ portion of APP lead to a higher ratio of Aβ1-42/Aβ1-40, or to the production of more total Aβ including the toxic 1-42 form. These mutations also produce the autosomal dominant form of AD.

Biochemical evidence suggests that the Aβ 1-42 form is the pathological species. It can aggregate to form small fibrils called oligomers, which are able to inhibit LTP and are toxic to neurons. It also appears that the oligomers can bind to a variety of synaptic surface receptors, probably preventing them from functioning normally. One of the receptors that Aβ binds to with high affinity is the prion protein. We will discuss this protein in a later section of this chapter.

The oligomers can also convert to amyloid fibers, which many scientists think are also toxic. However, some investigators believe that the formation of Aβ-containing amyloid plaques is a mechanism for removing the oligomers, thereby minimizing further damage to neurons.

One of the earliest changes seen in AD is the loss of synapses that correlates well with the first detection of memory deficits. Recently evidence has been presented that C1q and C3, two molecules released by microglia, bind to specific synaptic receptors. When this binding is accompanied by oligomer binding to putative receptors on the synaptic surface, there is resulting synaptic loss via engulfment by microglia. These effects may occur years before overt loss of neurons begins. (See General References.)

Another piece of evidence that indicates that Aβ plays a causal role in AD is the following. As discussed in the previous chapter, the APP gene is triplicated in patients with Down syndrome, generating more Aβ, ultimately causing AD.

Finally, a recent finding that strongly supports the amyloid hypothesis is the identification of a rare mutation in the Aβ sequence of APP. This mutation partially blocks the β-secretase, therefore favoring the action of the α-secretase that cleaves within the Aβ sequence. Individuals possessing this mutation have a decreased risk of contracting AD.

Some individuals have one or two copies of DNA that codes for the 4 forms of Apolipoprotein E (ApoE). Among its many putative functions in the brain, ApoE binds to Aβ and clears it from the brain. However, ApoE4 does not clear Aβ from the brain as well as the other allelic proteins, ApoE3 and ApoE2. Individuals with the ApoE4 allele do not invariably develop AD but their chances of developing Alzheimer's are 10−30 times those without it. These AD victims show symptoms beginning in their 60s, at least 10 years earlier than those with the idiopathic form (See General References.)

We might ask why the gene that codes for APP is not selected against, causing it to eventually disappear from the gene pool, especially since there are two homologous genes in the genome of vertebrates that lack the Aβ sequence and can apparently replace APP's known functions. This evidence, however, comes from studies in which the APP gene is deleted from laboratory mice and the mice appear to be healthy and breed normally. The simplest counter to this argument is that the experimental mice are raised in a totally artificial environment and if placed in the wild, they would not survive long enough to reproduce.

Even if the case for the need for APP could be made, there is still the problem caused by the Aβ sequence being embedded in the protein sequence of APP. As mentioned previously, the other two APP homologues do not possess this potentially deadly sequence. While no definitive evidence has been presented for a positive function of Aβ, there is evidence that in tissue slices made from mouse brain, Aβ serves to inhibit the excitatory effect of BDNF on neurotransmitter release similar to the role played by endocannabinoids. This may offer a reason for its remaining in the genome.

There is also evidence that the toxic Aβ 1-42 oligomers produce intracellular species of the tau protein that are hyperphosphorylated and aggregate into large NFTs. These tangles can block a variety of cellular functions. In fact, as discussed previously, tau pathology correlates much better with the origin and spread of the disease than does Aβ plaque pathology, leading some researchers to believe that it and not Aβ is the primary causative agent in AD.

Significant data implicate a trophic factor in the etiology of AD. A number of reports indicate that there is a surfeit of AD patients with the BDNF Val66Met variant compared with those who have the more common Val66Val isoform. As discussed in Chapter 17, Role of trophic factors and receptors in developmental brain disorders, these patients secrete less BDNF than do normal individuals.

Unfortunately the cause(s) of AD in the vast majority of cases has not been definitively identified, although it is apparent that lifestyle choices play a role. The risk factors for cardiovascular disease and AD are similar. People who eat a healthy diet and exercise are at decreased risk of developing AD. Also, people with higher education have a decreased risk.

Massive genetic and epidemiological investigations during the last 20 years have identified many gene variants that individually convey small additional risk for AD, but when combined with several others can produce a significant risk. One of the genes identified in this manner is the TREM2 gene. The TREM2 protein is on the plasma membrane of resident microglia. When the wild-type protein is present, the microglia form a protective coat around Aβ plaques, isolating them from surrounding neurons and preventing their toxicity. In contrast the mutant protein does not allow the microglia to form this protective coat. People with a mutation in TREM2 are at three times greater risk of contracting AD than are normal individuals. (See General References.)

There is a separation of many years between synaptic loss and neuronal loss in AD. This difference is found in the other connectional diseases. This time gap can be explained by the quantal hypothesis presented in Chapter 7, A testable theory to explain how the wiring of the brain occurs. According to this hypothesis, after initial synapses are lost, the affected neuron can survive at least partially by synaptic plasticity, which was described in Chapter 15, Synaptic pruning during development and throughout life depends on trophic factor interactions: the corollary described in Chapter 14, postulating the quantization of trophic factor requirements, provides a plausible explanation for the survival of neurons partially deprived of their full complement of trophic factors; however, its functionality is decreased. Eventually the number of trophic factor quanta received by the affected neuron diminishes to a level producing neuronal death. A cartoon illustrating this hypothesis is shown in Fig. 18.3A and B.

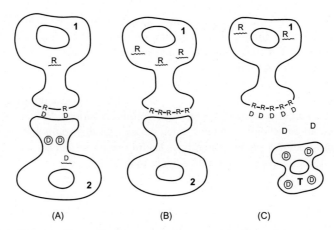

FIGURE 18.3 A cartoon illustrating the proposed manner by which a "global" adrenal medulla transplant can alleviate the symptoms of Parkinson's disease.

(A) This panel shows a substantia nigral dopaminergic neuron (2) synaptically and trophically connected to a striatal dopamine receptor-containing neuron (I). (B) Neuron 2 dies and neuron 1 produces many more dopamine receptors, thereby becoming supersensitive to dopamine. (C) The transplanted adrenal medullary chromaffin cell supplies a source of dopamine to allow the striatal cell to survive and remain functional.

Fine RE, Rubin JB. J Am Geriatr Soc. 1988;36:457–666. PMID: 2834427. Fig. 18.2. With permission.

Parkinson's disease (PD) is the second most common age-related neurodegenerative disease. It affects more than 1 million people in the United States alone and more than 53 million worldwide. With the increasing age of the population, the number of people with PD is expected to grow, although not as much as that of AD. The explanation for this fact is that in most affected individuals clinical signs of PD begin in the sixth or seventh decade of life. In contrast, the symptoms in the majority of AD cases begin in the late seventies or eighties.

The earliest symptoms of PD are related to movement and include slowing of gait and tremor. The disease then progresses to behavioral difficulties, including memory loss and diminished executive function, which occur in the late stages of the disease. PD is invariably fatal with death resulting approximately 10 years after the initial symptoms appear.

PD pathology first appears in the medulla and olfactory bulb. There is also evidence that there is early pathology in some neurons in the peripheral autonomic system. Clinical symptoms begin when pathology is seen in the synaptically connected midbrain nucleus, the substantia nigra. The small group of neurons that make up this nucleus is so named because the cells in this brain region contain neuromelanin, a black substance produced by the oxidation of metabolites of dopamine (See General References).

Dopamine is produced and secreted at presynaptic endings by the nigra neurons. When approximately 70% of the 100,000 dopaminergic neurons in the substantia nigra die, PD motor symptoms become evident. This is because the substantia nigra is a key nucleus in the circuit that modulates voluntary movement. Therefore it is to be expected that the motor symptoms characteristic of PD would begin here. It should be noted that the loss of about 70,000 of the 10^{12} neurons in the brain provides an excellent example that demonstrates that the death of a tiny fraction of the total neurons in the brain, if they are located in a vulnerable region, can lead to devastating consequences.

PD is the first example of a neurodegenerative disease involving a deficiency in a single neurotransmitter/trophic factor, dopamine. This discovery led to the use of the precursor of dopamine, L-dihyroxyphenylalanine (L-DOPA), which can readily cross the BBB, to treat patients successfully in the early stages of the diseases. For this discovery, Dr. Arvid Carlsson won the Nobel Prize in Physiology and Medicine in 2000. (See General References.) Also, in some patients dopamine receptor agonists are used to compensate for the loss of dopamine.

Unfortunately after several years of treatment the L-DOPA therapy becomes less effective, due to the continued loss of the dopaminergic cells. Many patients also develop tardive dyskinesia, a condition producing uncontrollable movements.

The evidence that the treatment of PD motor symptoms by L-DOPA and dopamine receptor agonists illustrates the key roles of receptors and ligands in PD. Dopamine would also be expected to act as trophic factor for regions innervated by the substantia nigra that are affected with the elimination of the dopamine-containing neurons as the disease progresses.

The major histopathological marker of PD is the cytoplasmic deposition of Lewy bodies and Lewy neurites. These are both insoluble amyloid aggregates composed mainly of a cytosolic protein, α-synuclein (Fig. 18.1, bottom left panel). In earlier stages of the disease the Lewy bodies are mainly found in the substantia nigra and other midbrain nuclei. At later stages, the Lewy bodies spread to connected cortical regions and in some cases produce dementia.

Lewy body disease is the second most common form of dementia and is difficult to distinguish clinically from AD. The disease is characterized by synuclein-containing aggregates, but they are found only in the cerebral cortex.

In the normal brain, synuclein is mainly found in presynaptic terminals where it associates with synaptic vesicles. One proposed function for synuclein is to maintain the normal supply of synaptic vesicles near the sites for neurotransmitter release. However, many other functions have been proposed. Just as the isolation of the Aβ-containing plaques in the brains of AD victims led to the hypothesis (still unproven) that Aβ is the cause of the disease, the identification of synuclein as the major constituent of Lewy bodies has led to the hypothesis that synuclein is the primary cause of PD and Lewy body disease. Supporting this hypothesis is the recent finding that synuclein has the ability to catalyze its own cleavage, and the resulting fragments are able to form amyloid. Also individuals with mutant forms of synuclein invariably develop an autosomal dominant form of PD, usually beginning in their forties.

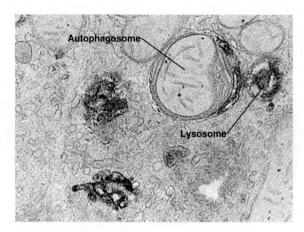

FIGURE 18.4 Electron micrograph.

From starved rat liver, an autophagosome containing a mitochondrion is in the process of fusing directly with a secondary lysosome.

Pollard TD, Earnshaw WC, Lippincott-Schwartz J, Johnson GT. Cell Biology. 3rd ed. Philadelphia, PA: Elsevier; 2017 [Chapters 1, 3, 13–16, 21–24, 27, 30, 33, and 34]. Fig. 34.12, p. 602. With permission.

Other mutations in proteins whose functions involve removal of damaged mitochondria, through a process related to autophagy called mitophagocytosis, also invariably cause PD (Fig. 18.4). Because the production of dopamine metabolites and their subsequent degradation involve oxidative reactions, generating free radicals that are likely to damage mitochondria, it is reasonable to expect that mutations in proteins involved in mitophagocytosis can cause PD. In fact two proteins, Parkin and Pink 1, that are necessary for the removal of nonfunctional mitochondria, produce autosomal dominant PD.

There is also an animal model indicating that damaged mitochondria play a key factor in PD. In the mid-1980s, a group of young individuals accidentally took a drug 1-methyl-4-phenyl-1,2,3,6-tetrahydropyridine (MPTP), thinking that it was cocaine. They quickly developed symptoms essentially identical to those of PD victims. In dopaminergic neurons only, MPTP is metabolized to 1-methyl-4-phenylpyridin-1-ium (MPP^+). MPP^+ then enters mitochondria and damages them. MPTP has been employed by many investigators as a valuable cell and animal model to investigate PD.

Recently a number of proteins whose primary roles are in various stages of endocytosis have been found to cause PD or PD-like syndromes when mutated. Among these are the retromer protein Vps35, which together with four other proteins forms a coated structure responsible for carrying recycled proteins including the Mannose-6-phosphate receptor from late endosomes to the *trans*-Golgi network. Other proteins that function in endocytosis include LARRK1 and auxilin. Some of these proteins have functions that implicate both mitochondria and organelles involved in endocytosis and protein degradation in the etiology of PD.

The taupathies are another group of connectional, age-associated neurodegenerative diseases. These include frontotemporal dementia (FTD), Pick's disease, and progressive supranuclear palsy. All of these diseases are characterized by the intracellular accumulation of hyperphosphorylated tau-containing NFTs, which form amyloid. Recently scientists have injected brain extracts from several different taupathies into the brains of transgenic mice expressing the human tau protein. The animals develop taupathies in the area of the brain from which each extract is prepared. For example, extracts from human brains with FTD develop tau inclusions in the frontal temporal regions. In contrast extracts from progressive supranuclear palsy brains produce tau aggregates in the thalamus and other midbrain regions while sparing the frontotemporal regions. The same results are produced when synthetic tau fibrils from brains of victims of these diseases are injected. Other studies indicate that fibers from each disease have a characteristic structure. These findings strongly suggest that the structure of the tau fibril may determine both the origin and the progression pattern of the disease. (See General References.)

Another disease that has received a great deal of interest during the last decade is chronic traumatic encephalopathy (CTE). This condition was first characterized in boxers who were hit in the head numerous times. Many years later they developed dementia commonly referred to as "punch drunkenness." Recently the same condition has been in found in football players and other athletes who have received numerous concussions, as well as in soldiers who have been exposed to several traumatic head injuries.

In autopsied cases, it appears that the pathology is wholly or mainly composed of tau-containing NFTs found in the depth of the sulci on the surface of the cerebral cortex. The NFT-containing lesion is initially confined to a small region usually near a blood vessel, but with time progresses to include many regions of the cerebral cortex and other parts of the brain. (See General References.)

There is evidence indicating that the first clinical signs of CTE occur at least 10 years after the traumas occur, in the form of depression and/or aggressive behavior. This stage is followed some years later by dementia that is similar clinically to AD.

An argument can now be made that CTE has the characteristics of a connectional disease. Its pathology consists of tau-containing NFTs. The pathology is localized at first to a small region of the brain and progresses with time to synaptically connected regions. Also the clinical symptoms begin, in many cases, well after the injury occurs. Then the symptoms gradually begin to appear and worsen over time.

Another age-related neurodegenerative disease that appears to be a connectional disease is Huntington's disease (HD). It is named for the doctor who first definitely described the symptoms and also demonstrated the inherited nature of the disease. HD affects about 4—15 Caucasians out of 100,000. The initial symptoms usually start in individuals between the age of 30 and 50.

HD is an autosomal dominant disorder that initially affects cells of the basal ganglia. The first clinical symptoms are subtle behavioral and/or cognitive changes, gradually leading to unsteady, jerky movements (chorea). This chorea led to its original name, Huntington's chorea. However, recently some patients have been shown to not develop chorea, so the name has been changed to HD. The disease eventually leads to the inability to talk, followed by dementia and death. The course of the disease from the time of initial diagnosis is 15−20 years, which is the pattern followed by other connectional diseases.

The molecular culprit in this disease is a protein called Huntingtin (HTT). HTT is found in all cells of the body but its expression is much higher in the brain. HD was the first autosomal disease discovered using genetic analysis of a large family with many members developing the disease. The chromosomal location of the HD disease gene was first determined using a technique called SNP analysis. A single-nucleotide polymorphism (SNP) is a variant in a single nucleotide that occurs at a specific position in the genome. At this site, each variant is present at a significant level within a given population (e.g., >1%). In the identical place in the human genome, C is found in most individuals, but the position is occupied by A in a small minority. SNP analysis was used with this large family, which had a high number of HD victims, to localize the mutant gene's position on the long arm of chromosome 4. This was followed by linkage analysis to define the exact chromosomal location in which the gene occurred, and finally by DNA sequencing to pinpoint the HTT gene and the genetic change leading to the disease. When an antibody to the mHTT was used for immune histochemistry it demonstrated a nuclear inclusion that contained the mHTT. (See General References.)

The mutation in HTT is also important in that it is one of the first diseases discovered that is produced by the repeat of a DNA triplet coding for an amino acid. In the case of HTT, it is CAG, which codes for glutamine (Q). Most people have 26 or fewer repeats of HTT. However, when a person's HTT gene contains 40 or more repeats, the affected individual will certainly develop the disease. As mentioned previously, HD is an autosomal dominant disease with the affected gene coming from either parent. It therefore appears that the mutated HTT produces a toxic gain-of-function since one copy of the normal gene is still present.

Mutant HTT (mHTT) interacts with numerous proteins that influence a variety of important processes. Of particular relevance to our hypothesis concerning the important role of receptors and trophic factors in age-related neurodegenerative diseases is the fact that mHTT's extra Q residues can interact with Qs on the important transcriptional coactivator, CBP. This in turn causes the decreased transcription of many proteins that are important for neuronal function. One of the affected proteins is BDNF, which has numerous functions in the brain as discussed in Chapter 10, The key roles of BDNF and endocannabinoids at various stages of brain development including neuronal commitment, migration and synaptogenesis. There is also evidence that mHTT can inhibit neurotransmitter release, although the precise mechanism is still not well understood. Finally, fragments of mHTT can form intracellular amyloid, trapping other molecules within its insoluble meshwork. Needless to say, this amyloid can disrupt many cellular processes. However, many scientists argue that the amyloid protects the neurons by removing the much more toxic soluble mHTT.

ALS, also known as Lou Gehrig's disease or motor neuron disease, results from the death of motor neurons innervating both the upper and lower body muscles. It is a rare disease affecting about 1—2 per 100,000 individuals. Approximately 90% of individuals with the disease have no known cause, while the other 10% have a gene defect.

ALS is thought to gradually affect the motor neurons with symptoms becoming detectable only after about 10—20 years, similar to the course of other connectional diseases. The first symptoms are usually muscle stiffness and twitching followed by loss of speech, inability to swallow, and eventually death. The course of the disease after diagnosis is usually quite rapid, 3—4 years. There are exceptions, for example, Stephen Hawking, the great physicist who survived over 50 years after diagnosis.

A protein whose mutation is linked to almost half of genetically caused ALS cases as well as to FTD is C9ORF72. The function(s) of this protein are not well understood, but it is thought to play a role in the processes of endocytosis and autophagy.

Most individuals have a few copies of the hexanucleotide repeat GGGGCC in the noncoding region of C9ORF72. However, individuals who have ALS have hundreds of repeats. One theory about the way in which the extra repeats cause the disease is that the mutation interferes with the normal expression of the mRNA coding for the protein. Another theory postulates that the possession of only half of the normal quantity of the protein causes the disease. Finally it is known that the hexanucleotide repeats can be translated into dipeptides that can be toxic to the cells (Fig. 18.5).

Another protein that is responsible for about 20% of the individuals with the genetic form of the disease is superoxide dismutase, a cytoplasmic enzyme that destroys the toxic superoxide free radical constantly produced by mitochondria. Mutations in the gene coding for the enzyme, of which there are more than 50, can result in either a loss or gain in function. A loss can result in the inability of the neuron to remove the toxic superoxide radical. However, some scientists suggest that a gain in a toxic function results from a misfolding of the molecule, producing an amyloid species that can affect many cellular processes. While the protein is ubiquitous, the disease only affects the regions of the brain mentioned previously, while sparing others.

Two other proteins have more recently been linked to ALS. TBP-43 and FUS are normally both nuclear mRNA binding proteins that help to transport the mRNAs to the cytoplasm. When the genes coding for these two are mutated, ALS results. TBP and FUS are also found in the amyloid-containing aggregates seen in autopsied brains of patients with the sporadic form of the disease. Animal models

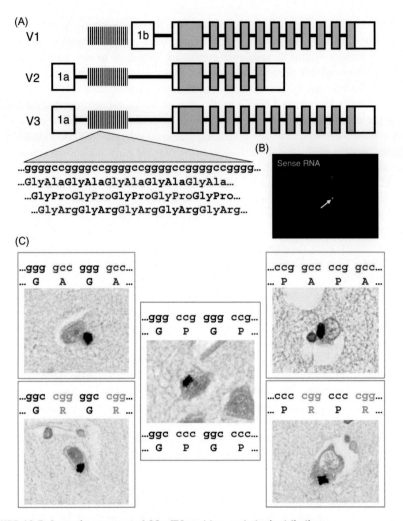

(A)

V1

V2   1a

V3   1a

...ggggccggggccggggccggggccggggccgggg...
...GlyAlaGlyAlaGlyAlaGlyAlaGlyAla...
...GlyProGlyProGlyProGlyProGlyPro...
...GlyArgGlyArgGlyArgGlyArgGlyArg...

(B) Sense RNA

(C)

...ggg gcc ggg gcc...
... G   A   G   A ...

...ccg gcc ccg gcc...
... P   A   P   A ...

...ggg ccg ggg ccg...
... G   P   G   P ...

...ggc cgg ggc cgg...
... G   R   G   R ...

...ccc cgg ccc cgg...
... P   R   P   R ...

...ggc ccc ggc ccc...
... G   P   G   P ...

FIGURE 18.5 Genomic structure of C9orf72 and key pathological findings.

(A) Depiction of the three main transcript of C9orf72. The GGGGCC-repeat expansion (about 100–5000 repeats compared with <30 repeat in controls) is either located upstream of the coding region (*orange*), in the promoter region (isoform V1) or in the first intron (isoforms V2 and V3). V1 and V3 encode the same ~54 kDa isoform, while V2 encodes a C-terminally truncated ~25 kDa isoform. The sense transcript (ggggcc RNA repeat in lower case) is translated in all three reading frames into abundant DPR proteins. (B) RNA foci formed by the sense strand (arrow) are detectable by in situ hybridization in the nucleus. (C) Specific antibodies detect DPR inclusions derived from non-ATG translation of sense and antisense repeat transcripts in all reading frames.

Edbauer D, Haass C. An amyloid-like cascade hypothesis for C9orf72 ALS/FTD. Curr Opin Neurobiol. 2016;36:99 – 106. PMID: 26555807. Fig. 18.1. With permission.

of mice over- and underexpressing both molecules suggest either loss-of-function or toxic gain-of-function mechanisms can occur.

The neurotransmitter GLU, which can also serve as a trophic factor, has also been implicated in ALS. Excess GLU can produce excitotoxicity by opening the channel in the NMDAR, thereby allowing too much Ca^{++} into the postsynaptic terminals. The excess cytoplasmic Ca^{++} then activates many enzymes, triggering a host of events leading to neuron death. Patients with ALS are found to have more glutamate in their cerebrospinal fluid and blood than do normals. Based these findings, ridazole, a glutamate receptor antagonist, has been approved for the treatment of ALS, although it only prolongs life for 2–4 months. Obviously there are many other pathologic mechanisms involved, some of which have been discussed in this chapter.

In all the age-related connectional diseases described in this chapter, an amyloid species is found to be present, which in some cases appears to spread synaptically. In the affected neurons, protein/proteins that are major constituents of these amyloid fibers also cause the disease when mutated.

Very interesting models for these types of connectional diseases are the prion diseases. Kuru, a human prion disease, was first described in a tribe in New Guinea who ate the brains of dead relatives as a mark of esteem. The prions in the dead brain were conveyed to the family who ate the brains. It takes approximately 20 years after prion ingestion for the clinical symptoms to begin appearing.

After the kuru victims died, autopsies demonstrated that their brains were characterized by four lesions. These include neuronal loss, proliferation of astrocytes, amyloid plaques, and most characteristically, the sponge-like appearance of the brain caused by the massive loss of neurons in the brain. This spongiform appearance is characteristic of all prion diseases (Fig. 18.6).

The scientist who first described kuru, Dr. Carleton Gadjusek, was awarded the Nobel Prize in Physiology and Medicine in 1976 for this discovery. (See General References.) He tried for many years to demonstrate that the infectious agent of kuru is a virus, or at least contained either DNA or RNA. However, he was unable to demonstrate that a nucleic acid caused the disease.

In the 1980s, Stanley Prusiner isolated an integral plasma membrane sialoglycoprotein receptor that he called the prion. His work was at first met by great skepticism because he postulated that a protein could cause a disease without the presence of a nucleic acid, an idea that violated the central dogma of biology; namely, that only nucleic acids can be self-replicating molecules. However, after a series of elegant experiments, he convinced the skeptics. For this work, Dr. Prusiner was awarded the Nobel Prize in Physiology and Medicine in 1997. (See General References.)

The infectious prion, which in some but not all cases is resistant to a powerful protease, protease K, is not inactivated by any cleaning agent except bleach. This means that investigators have to take extreme precautions when working with prions, both in animal and human brains and with tissue cultured cells (Fig. 18.6).

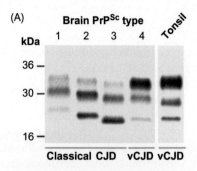

(A) **Brain PrP^{Sc} type**

kDa

36 —

30 —

16 —

Classical CJD   vCJD   vCJD

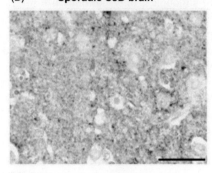

(B)   **Sporadic CJD brain**

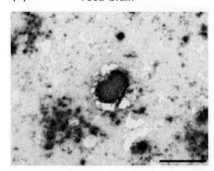

(C)   **vCJD brain**

FIGURE 18.6 Characterization of disease-related prion protein in human prion disease.

(A) Immunoblots of proteinase K digested tissue homogenate with anti-PrP monoclonal antibody 3F4 showing PrPSc types 1–4 in human brain and PrPSc type 4t in vCJD tonsil. Types 1–3 PrPSc are seen in the brain of classical forms of CJD (either sporadic or iatrogenic CJD), while type 4 PrPSc and type 4t PrPSc are uniquely seen in vCJD brain or tonsil, respectively. (B and C) Brain from patients with sporadic CJD or vCJD show abnormal PrP immunoreactivity following immunohistochemistry using anti-PrP monoclonal antibody ICSM35. Abnormal PrP deposition in sporadic CJD brain most commonly presents as diffuse, synaptic staining, whereas vCJD brain is distinguished by the presence of florid PrP plaques consisting of a round amyloid core of PrP surrounded by a ring of spongiform vacuoles. Scale bars B and C = 50 μm.

Wadsworth JD, Collinge J. Update on human prion disease. Biochim Biophys Acta. 2007;1772:5982609. PMID: 17408929. Fig. 1. With permission.

The normal prion becomes infectious when it encounters a molecule of the infectious prion. Then the normal protein changes its conformation to the structure of an amyloid fiber. (See General References.) This structural change requires the presence of glycoaminoglycans for the conversion to infectivity to occur.

There are many known prion diseases in mammals. Many of these diseases are caused by ingestion of the infectious agent, including kuru and scrapie, a disease of sheep. Scrapie spreads through a colony of sheep by accidental ingestion of grass containing prion-infected urine. Spongiform bovine encephalopathy (also known as mad cow disease) can be spread by cattle eating the ground-up remains of infected cattle. This disease can also spread to humans who eat meat from the infected animals. As of 2014, 222 people who ate meat from infected cattle have contracted the disease. This unfortunate situation is a stark demonstration that prions are able to cross the species barrier and infect humans.

The infectious prion is not only found in the brain but in other tissues of the body, as is also shown histochemically by use of a specific antibody against the prion. In fact the prion protein is ubiquitous in the body of an infected individual, even though its highest concentration is found in the brain.

An important finding was made using brains from cattle killed at various stages of mad cow disease. The symptoms begin subtly, with early clinical symptoms being an abnormal gait, and behavioral changes that include tremors and heightened sensitivity to a variety of stimuli. Later the animal's gait is more severely affected due to the loss of muscle control. Aggression, anxiety, and nervousness may also occur at this stage of the disease. Once clinical symptoms arise, they invariably worsen over the upcoming months, inevitably producing death.

In other prion diseases, including those in humans, the symptoms begin with subtle mood changes and gait disturbances. As the disease progresses, dementia and death ensue. There are also rare genetic forms of human prion disease, including Creutzfeldt—Jakob disease and fatal familial insomnia. These diseases are both caused by mutations in the gene coding for the prion protein.

As was definitely shown in cattle infected with mad cow disease, the early symptoms of the prion diseases begin after a prolonged incubation period, and proceed inexorably to dementia and death. Therefore they meet many of the criteria of connectional diseases. However, to date, no single brain region has been identified as the origin of any of the prion diseases, in contrast to the other connectional diseases mentioned in this chapter.

The prion is believed to normally function as a receptor. There is data suggesting that the normal prion serves to transport protein ligands into cells. Among the ligands to which it binds is Aβ. The prion binds more strongly to the toxic, oligomeric form, thereby potentially playing a role in AD.

Many investigators, including Dr. Gadjusek, have tried to infect chimpanzee brains with brain extracts from AD patients with no success. However, investigators have injected brain extracts from mice carrying the human APP gene injected into other human APP—carrying mice. These mice developed amyloid deposits in regions of the brain affected in AD victims as well as a deterioration in memory.

The jury is still out on whether or not the proteins thought to cause AD, PD, HD, ALS, and taupathies can also be infectious, as are prions. They do, however, have a striking resemblance to the prion diseases in that they are all capable of producing amyloid, and the infectious protein appears to serve as a "seed" to convert the normal protein into an amyloid protein. There are also both genetic and sporadic forms of the abovementioned diseases. Also they all are connectional, spreading slowly from one region of the brain to another, synaptically connected region.

Recently researchers have demonstrated that in several of the age-related neurodegenerative diseases, proteins thought to play key roles in pathogenesis can be released from one cell and taken up by another cell that is synaptically connected. Among the proteins that appear to be transmitted in this fashion are the tau protein and synuclein. It is still not clear how these cytoplasmic molecules can pass through the plasma membrane of one neuron and enter another. One possibility is that they have a confirmation that allows them to grow directly through the plasma membrane. Another possibility is that they are encapsulated in exosomes, which are fragments of the plasma membrane that bud off into the extracellular space, while containing a small portion of the cytoplasm. There is also the possibility that microglia are involved in the transfer of the pathogenic species.

While prions are mainly studied as infectious agents in the brain and other organs, there is evidence that at least some prions may serve useful purposes. An example is the neuronal synaptic protein, cytoplasmic polyadenylation binding protein (CPEB). This protein, when in its monomeric form, has a very short half-life. However, when it aggregates to a multimeric form that assumes the amyloid structure, it has a very long half-life. (See General References.) This is a useful property for a protein that serves as a molecular component, maintaining a specific long-term memory in a particular neuron or group of neurons. As its name implies, CPEB in synaptic endings binds to the polyadenylation sites of a specific group of mRNAs, thereby stimulating their translation into proteins.

At present the amyloid form of CPEB is the only mammalian protein shown to play a useful role in brain function. However, many cellular proteins can form amyloid. It would therefore be expected that other potentially amyloid-forming proteins will be found to play useful roles in memory and other functions.

GENERAL REFERENCES

Book Chapters

Kandel ER, Schwartz JH, Jessel TM, Siegelbaum SA, Hudspeth AJ. *Principles of Neural Science*. 5th ed. New York: McGraw Hill; 2013 [Chapters 53, 56, 66, 67].

Squire LR, Berg D, Bloom FB, Du Lac S, Ghosh A, Spitzer NC. *Fundamental Neuroscience*. 4th ed. Oxford, UK: Elsevier; 2013 [Chapters 28, 30 43, 48].

Review Articles

Appel SH. A unifying hypothesis for the cause of amyotrophic lateral sclerosis, parkinsonism, and Alzheimer's disease. *Ann Neurol*. 1981;10:449−505.

Benes FM. Carlsson and the discovery of dopamine. *Trends Pharmacol Sci*. 2001;22:46−47.

Esposito L, Pedone C, Vitagliano L. Molecular dynamics analyses of cross-beta-spine steric zipper models: beta-sheet twisting and aggregation. *Proc Natl Acad Sci USA*. 2006;103:11533−11538.

Iba M, Guo JL, McBride JD, Zhang B, Trojanowski JQ, Lee VM. Synthetic tau fibrils mediate transmission of neurofibrillary tangles in a transgenic mouse model of Alzheimer's-like tauopathy. *J Neurosci*. 2013;33:1024−1037.

Jiang P, Dickson DW. Parkinson's disease: experimental models and reality. *Acta Neuropathol*. 2018;135:13−32.

Jiang T, Yu JT, Zhu XC, Tan L. TREM2 in Alzheimer's disease. *Mol Neurobiol*. 2013;48:180−185.

Kumar DK, Choi SH, Washicosky KJ, et al. Amyloid-β peptide protects against microbial infection in mouse and worm models of Alzheimer's disease. *Sci Transl Med*. 2016;8:340−372.

Netzer WJ, Bettayeb K, Sinha SC, Flajolet M, Greengard P, Bustos V. Gleevec shifts APP processing from a β-cleavage to a nonamyloidogenic cleavage. *Proc Natl Acad Sci USA*. 2017;114:1389−1394.

Poirier J. Apolipoprotein E and Alzheimer's disease. A role in amyloid catabolism. *Ann NY Acad Sci*. 2000;924:81−90.

Prusiner SB. Prions. *Proc Natl Acad Sci USA*. 1998;95:13363−13383.

Rayman JB, Kandel ER. Functional prions in the brain. *Cold Spring Harb Perspect Biol*. 2017;9:a023671.

Venkitaramani DV, Chin J, Netzer WJ, et al. Beta-amyloid modulation of synaptic transmission and plasticity. *J Neurosci*. 2007;27:11832−11837.

Zetterström R. The discovery of misfolded prions as an infectious agent. *Acta Paediatr*. 2010;99:1910−1913.

Neuronal stem cells

19

CHAPTER OUTLINE

The theory presented in Chapter 7, A testable theory to explain how the wiring of the brain occurs, states that specific trophic factor—receptor interactions are crucial in creating the circuitry of the brain. A hypothesis can now be offered based on this theory. It states that if a particular neuron is deprived of a synaptically connected partner and therefore its trophic support, the neuron will send out the complementary trophic factor(s) to attract another partner. In other words, a neuronal precursor cell exists that has the complementary array of receptors to recognize the trophically and presumably tropic signals the deprived neuron is sending (Fig. 19.1).

There is substantial evidence to support this hypothesis. An example is an elegant set of experiments by Sotelo and associates performed in the pcd mouse. In this mouse line the Purkinje neurons in the cerebellum develop normally and then begin to die 2 months after birth. However, if a graft of fetal cerebellum is placed over the cerebellar surface of a 4-month-old pcd mouse, the Purkinje cell precursors and no other cell type in the graft will migrate into the interior of the cerebellum and completely replace the dead Purkinje cells. The graft must contain 12-day-old fetal cells. If younger or older grafts are used, no Purkinje cells or any other neurons leave the graft. This experiment emphasizes the critical importance of using Purkinje cell precursors that contain the correct receptors to respond to the signals from the disconnected neurons. Also these receptors are apparently only transiently expressed.

The transplanted neurons make synapses with the appropriate neurons, as shown by both anatomical and physiological experiments. Unfortunately, the cells cannot grow axons long enough to reach their appropriate targets. This is likely because the myelin surrounding already present Purkinje axons contains proteoglycans, which form a barrier for further growth of the transplanted Purkinje cell axons. This barrier to central nervous system axonal elongation does not occur in the peripheral nervous system, presumably because the peripheral myelin does not contain the same proteoglycans on its surface (Fig. 19.2).

More recently, human cells from the embryonic ganglionic eminences containing GABAergic neuron precursors have been injected into mouse brains and

Receptors in the Evolution and Development of the Brain. DOI: https://doi.org/10.1016/B978-0-12-811012-6.00019-4

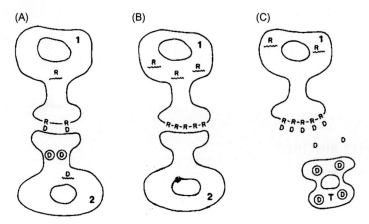

FIGURE 19.1 A cartoon illustrating the proposed manner by which a "global" medulla transplant, as defined by Sotelo and Alvardo-Mallart, could alleviate the symptoms of Parkinson's disease.

(A) This panel shows a substantia nigral dopaminergic neuron (2), synaptically and tropically connected to a dopamine receptor (R) containing neuron (1). (B) Neuron 2 dies and neuron 1 produces many more dopamine receptors, thereby becoming supersensitive to dopamine (D). (C) The transplanted adrenal medullary chromaffin cell (T) supplies a source of dopamine to allow the striatal cell to survive and remain functional.

Fine R, Rubin J. Specific trophic factor-receptor interactions: key selective elements in brain development and aging. J Am Geriatr Soc. *1988;36:457–466. PMID: 2834207. Fig. 2. With permission.*

appear to structurally and functionally integrate into existing circuitry (Fig. 19.3). Another remarkable finding is that the transplanted GABA neuronal precursors follow their intrinsically programmed developmental course with no alterations produced by introduction into the already mature brain (Fig. 19.4).

As was discussed in Chapter 8, The development of the cerebral cortex, GABAergic interneuron precursors migrate tangentially long distances during their development, survive in very different environments than those in which they originated, and integrate into already formed excitatory neuronal circuits. This may be why the transplanted GABAergic precursors have a unique ability to replace dying neurons in the cerebral cortex and other areas of the brain. This has been demonstrated in several animal models of neurological and psychiatric diseases including epilepsy, schizophrenia, chronic pain, and Parkinson's disease. (See General References.)

GABAergic precursors cells can be injected into normal mice and form additional functional synapses. These injected mice score higher on several cognitive tests than do the normal mice. One possible reason for this result is that the normal cerebral cortex does not possess enough GABAergic neurons to totally saturate the postsynaptic endings of the excitatory projection neurons. This may result from the high number of apoptotic GABAergic precursors during their prolonged

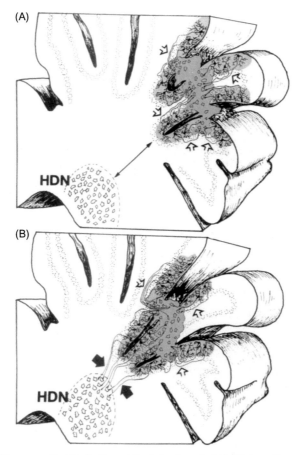

(A)

(B)

HDN

HDN

FIGURE 19.2 Diagrammatic illustration of the fate of grafted Purkinje cell axons that succeed in leaving the host molecular layer.

The graft remnant (*blue area*) containing grafted deep nuclear neurons remains within the host parenchyma. Grafted Purkinje cells occupy, for variable distances (*red area*), the host molecular layer. (A) The distance between grafted Purkinje cells and the host deep nucleus (HDN) is greater than 600 µm. The Purkinje cell axons do not orient toward their target (open arrows) but penetrate the graft remnant and establish synaptic contacts with grafted deep nuclear neurons. (B) The distance between some of the grafted Purkinje cells and the HDN is less than 600 µm. The axons of the nearest Purkinje cells cross the host white matter (solid arrows) and penetrate the HDN, where they make synaptic contacts on their specific targets, partially reconstructing the corticonuclear projection. However, most of the axons of the distant Purkinje cells (open arrows) terminate within the graft remnant. This drawing illustrates the hypothesis that competing gradients of chemoattractant molecules are produced by deep nuclear neurons from both the solid graft and the host. Such molecules produced by the host may orient the navigation of the growth cones of grafted Purkinje cell axons, once they succeed in crossing the host granular layer.

Sotelo C, Alvarado-Mallart RM. *The reconstruction of cerebellar circuits.* Trends Neurosci. 1991;14:350–355. PMID: 1721740. Fig. 5. With permission.

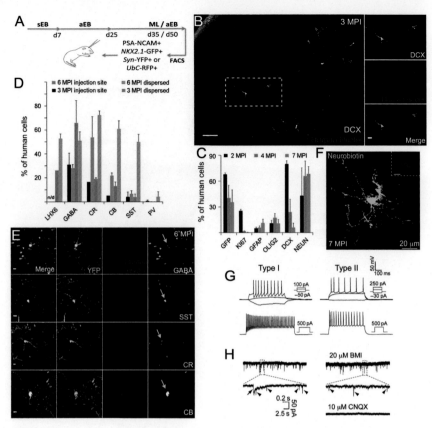

FIGURE 19.3 hPSC-derived interneuron subtype maturation and functional integration in the mouse brain.

(A) hPSC-derived MGE-like interneuron precursors FACS-sorted for NKX2.1-GFP and PSA-NCAM, labeled with YFP or RFP virus, and injected into newborn mouse cortex. (B) Human nuclei + cells expressed DCX and migrated into the cortex by 3 months postinjection (MPI). Blue: DAPI. Scale bar: 100 μm. Right panel is zoom of dashed rectangle in separate channels and merged. Scale bar: 20 μm. (C and D) Quantification of lineage-specific marker histology at 2 (*black*), 4 (*orange*), and 7 (*blue*) MPI with d35 ML cells (C), or of subtype markers at 3 (*black, blue*) and 6 (*orange, green*) MPI with d50 aEB cells. (D) Plotted as human cells near or dispersed from the injection site. Data represented as mean ± SEM. (E) Histological analysis of human nuclei + cells prelabeled with Syn-YFP at 6 MPI that coexpressed (arrow) subtype markers. Blue: DAPI. Scale bar: 20 μm. (F) hPSC-derived neuron labeled by intracellular filling of neurobiotin (NB, *green*). Inset: UbC-RFP fluorescence of filled neuron 7 MPI. Scale bar: 20 μm; inset 5 μm. (G) Traces of AP firing patterns of type I (left) and type II (right) hPSC-derived neurons upon near-threshold (top) and super-threshold (bottom) current injection at 7 MPI. Scale bars: 50 mV and 100 ms. (H) Left panel: traces of spontaneous PSCs recorded from hPSC-derived neurons at 7 MPI; upper right: BMI blocked PSCs with slow decay-time (arrow), and the remaining PSCs with fast decay-time (arrow head) were blocked by subsequent application of CNQX (lower right panel). Scale bars: 50 pA, 2.5 s, and 0.2 s (dashed line) for zoomed traces.

Nicholas CR, Chen J, Tang Y, et al. Functional maturation of hPSC-derived forebrain interneurons requires an extended timeline and mimics human neural development. Cell Stem Cell. 2013;12:573–586. PMID: 23642366. Fig. 7. With permission.

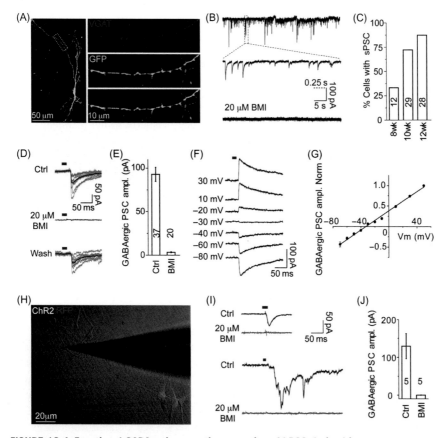

FIGURE 19.4 Functional GABAergic synaptic properties of hPSC-derived interneurons.

(A) Images showing VGAT expression in hPSC-derived NKX2.1-GFP + neurons at 12 WPD. Right: zoom of dashed rectangle. Scale bar: left, 50 μm; right, 10 μm. (B) Traces showing spontaneous postsynaptic currents (PSCs) in hPSC-derived neurons, bottom: PSCs were fully blocked by BMI. Scale bar: 100 pA, 5 s, and 0.25 s (dashed line) for middle trace. (C) Percentage of neurons showing spontaneous PSCs at different stages. (D) hPSC-derived neurons were transfected with ChR2-EYFP. Traces show pulses of blue light (*blue bar*) evoked PSCs in neighboring cells that were reversibly blocked by BMI. Scale bar: 50 pA and 50 ms. (E) Average amplitudes of light-evoked GABAergic PSCs and application of BMI. (F and G) Traces showing light-evoked (*blue bar*) PSCs at different holding potentials. Summarized results (n = 7) showing I-V curve of light-evoked GABAergic PSCs (G). (H) Merged image showing DIC of human fetal cortical cells cocultured with sorted UbC-RFP + and ChR2 transfected hPSC-derived neurons. Scale bar: 20 μm. (I) Traces showing blue light (*blue bar*) stimulation of hPSC-derived neuron-evoked PSCs in RFP-negative recorded human fetal cortical neurons. Upper panel shows PSC monosynaptic response, lower panel shows PSC with polysynaptic responses—both fully blocked by BMI. Scale bar: 50 pA and 50 ms. (J) Averaged amplitudes of light-evoked PSCs and application of BMI. (E and J) Data represented as mean ± SEM.

Nicholas CR, Chen J, Tang Y, et al. Functional maturation of hPSC-derived forebrain interneurons requires an extended timeline and mimics human neural development. Cell Stem Cell. *2013;12:573–586. PMID: 23642366. Fig. 6. With permission.*

migration period. Therefore the excitatory neurons that have not received a saturating supply are continuously signaling for additional GABAergic neurons. (See General References.)

Another remarkable experimental finding is that GABAergic precursor injection can reopen the window of ocular dominance plasticity in the recipient visual cortex. This result occurs irrespective of the age of the recipient mouse. (See General References.)

There have also been many experiments using mouse models of Parkinson's disease (PD), in which specifically dopaminergic neurons in the substantia nigra die during aging and can be at least partially replaced by fetal dopaminergic neurons. There have even been experimental procedures involving injection of human fetal neurons into the brains of humans with PD. These have met with limited success, probably in part because these cells are attacked by microglia that regard the injected neurons as foreign. Another possible reason for the limited success of these transplants is that the harvested fetal cells may not be at the precise stage of development in which they express the correct set of receptors. This prevents them from responding correctly to the trophic and tropic signals produced by the disconnected neurons. (See General References.)

A major breakthrough was made several years ago by Dr. Shinya Yamanaka. He transduced four human oncogenes (cancer-causing genes) into differentiated cells, for example, white blood cells or skin fibroblasts. This injection caused them to dedifferentiate and become totipotent stem cells, capable of differentiating into any cell type in the body. Dr. Yamanaka won the Nobel Prize in Medicine and Physiology in 2012 for this discovery. (See General References.)

Already this finding has had a profound effect on the study of human diseases including the age-related neurological connectional diseases described previously. For example, one can take white blood cells from Alzheimer's disease (AD) patients, turn them into stem cells and then into a variety of neuronal precursor cells. The investigator can grow these cells in cell culture and study their properties, then compare them with those derived from the white blood cells of normal, age-matched individuals. The AD patient—derived cells can also be used to test various drugs that may normalize the cells.

In the future, the hope is that normal neuronal precursor cells can be injected into the damaged region of the brain of a patient with a neurodegenerative disease whose trophically and synaptically disconnected neurons are signaling the precursors, as shown in Fig. 19.1. The injected cells can then migrate to the affected sites and replace the dying neurons. A major advantage of the use of neuronal precursors derived from the cells of the patient is that there will be no problems associated with cells of the immune system attacking the injected precursors. However, the disadvantage of using the patient's own cells is that they have the exact genotype as the patient. Hence they carry genes that can cause the disease. This fact should not be a problem in age-related diseases such as AD and PD. In these diseases, the phenotype is expressed many years after birth. Therefore, the neuronal precursor—injected patients will most certainly die before the injected neurons start to sicken. Also, in diseases that have at least a partially environmental cause, the precursors may not be affected.

There are already ongoing animal studies to test the possibility of using human stem cell—derived neuronal precursors to treat diseases. These include studies using substantia nigra—derived dopaminergic precursor cells and GABAergic interneuron precursor cells.

Dopaminergic precursor cells have been injected into the substantia nigras of monkeys after MPTP treatment. These precursors can replace the dead neurons, as indicated by the immunostaining with tyrosine hydroxylase (TH) a specific marker for dopamine-containing neurons. (See General References.) These dopaminergic precursors offer the distinct possibility of treating PD since only one specific neuronal cell type needs to be replaced. Also, as discussed in Chapter 18, Trophic factor-receptor interactions which mediate neuronal survival can in some cases last through life, and trophic deficits can produce connectional neurodegenerative diseases, only a partial replacement is necessary. This is because individuals with more than 30% of their complement of dopaminergic neurons can function normally.

GABAergic neuronal precursors produced from the reprogramming of mouse or human stem cells have been injected into damaged areas of mouse brains and appear to migrate and partially change into mature GABAergic neurons. Unfortunately, the cell culture conditions necessary for the production of neuronal precursor cells able to become the specific subtypes of mature GABA interneurons have not yet been achieved.

As previously mentioned, because many human diseases including epilepsy, schizophrenia, and autism have a deficit in GABAergic neurons, precursor injections may ultimately offer a treatment for these devastating diseases. Also as mentioned previously, GABAergic neuronal containing transplants injected into normal mouse brains produce cognitively superior animals. Therefore, the possibility arises of using GABAergic precursor cells prepared from human stem cells injected into normal human brains to increase their intellectual capabilities. Obviously these experiments would face many scientific and ethical hurdles before they could be attempted.

GENERAL REFERENCES
Recommended Book Chapters

Kandel ER, Schwartz JH, Jessel TM, Siegelbaum SA, Hudspeth AJ. *Principles of Neural Science*. 5th ed. New York: McGraw Hill; 2013.

Squire LR, Berg D, Bloom FB, Du Lac S, Ghosh A, Spitzer NC. *Fundamental Neuroscience*, 4th ed. Oxford, UK: Elsevier; 2013 [Chapter 14].

Recommended Review Articles

Daley GQ. Cellular alchemy and the golden age of reprogramming. *Cell.* 2012;151:1151–1154.

Fine R, Rubin J. Specific trophic factor-receptor interactions: key selective elements in brain development and aging. *J Am Ger Soc.* 1988;36:457–466.

Olson L. Regeneration in the adult central nervous system: experimental repair strategies. *Nat Med.* 1997;3(12):1329–1335.

Sotelo C, Alvarado-Mallart RM. The reconstruction of cerebellar circuits. *Trends Neurosci.* 1991;14:350–355.

Southwell DG, Nicholas CR, Basbaum AI, et al. Interneurons from embryonic development to cell-based therapy. *Science.* 2014;344:1240622.

Summary and conclusions 20

CHAPTER OUTLINE

20.1 SUMMARY

After defining the properties including specificity, selectivity, and the ability to be regulated, which make the receptors and their ligand/trophic factors critical for the evolution and development of the "big" brain, we discuss categories of receptors. The most abundant of these are the G protein coupled receptors (GPCRs), which together comprise over 10% of the human exome (the portion of the genome coding for proteins). Other families of receptors include the tyrosine and serine/threonine kinases; ionic receptors, including glutamate and GABA receptors, nuclear receptors; and nutritional receptors.

Once we have categorized these receptors, the signaling pathways used to transmit information are described. These include the GPCR signaling pathways that utilize a variety of G proteins, protein phosphorylation activating a variety of protein kinases and phosphatases, and release of binding proteins leading to gene transcription.

We next describe mechanisms of up- and downregulation of receptor number to increase or decrease the sensitivity of a particular receptor to the concentration of a given ligand. We all have experienced the strong odor produced by steamed broccoli. However, in a very short period time the odor decreases markedly. This is due to the downregulation of the odorant receptors on nasal epithelial cells that detect the odor of steamed broccoli. In contrast, bloodhounds tracking the scent of an escaped prisoner increase the number of odorant receptors for smells identical to those emanating from the prisoner's bedsheets.

Finally, the synthesis and transport of receptors from the endoplasmic reticulum to the plasma membrane is detailed. The amount of a particular receptor synthesized is also controlled by the concentration of a given ligand.

Once the various properties of receptors, their signaling modalities, and synthetic pathways are described, we begin a journey starting with the evolution of the most primitive organisms, that is, members of the phylum Archaea, from

Receptors in the Evolution and Development of the Brain. DOI: https://doi.org/10.1016/B978-0-12-811012-6.00020-0

245

which all other species descend. Regarding one member of Archaea, two key receptors on the blue algae's plasma membrane, that is, Bacteriorhodopsin and Halorhodopsin, carry out two essential processes necessary for life. Bacteriorhodopsin uses the energy obtained by the removal of H^+ from the cell to synthesize ATP, the universal energy-carrying molecule. Halorhodopsin can sense the direction of light causing the algae to move toward its source, the sun (phototaxis), enabling it to obtain the nutrients it requires.

Besides studying its functional properties, Bacteriorhodopsin's three-dimensional structure has been determined to 3-angstrom resolution. This makes it the first integral membrane protein whose structure has been determined. By serendipity its structure turned out to be very similar to those of the family of GPCRs, even though their peptide sequences are very different. One explanation for this finding is that because the structural properties of Bacteriorhodopsin are so valuable, they evolved independently in more complex organisms.

Bacteria are the most numerous phylum on Earth. They have employed receptors for numerous functions. Many bacterial species have flagella that beat, propelling the organism through water. By use of signaling involving histidine and tyrosine kinases, the bacterium can reverse the direction of flagellar beating, propelling the organism toward a source of food or away from a predator.

While bacteria are thought of as single-celled organisms, many species possess receptor-mediated quorum sensing, which allows them to determine the concentration of other members of their species. These bacteria can then form a biofilm, which functions in many ways like a multicellular organism. This biofilm synthesizes different classes of molecules and the bacteria become resistant to toxins, including antibiotics.

Choanoflagellates are thought to be the nearest protist relative of all multicellular animals. They have a long flagellum that, by beating, allows the organism to seek food and flee predators. The sequenced genome reveals that Choanoflagellates have many genes that code for receptors, including 10 GPCR genes. In contrast, sponges, one of the most primitive multicellular eukaryotes, have over 500 GPCR genes, indicating a great expansion of this very important receptor family when multicellularity evolved.

The slime mold *Dictyostelium discoidium* represents a transitional eukaryote, transforming from a single-celled organism to a multicellular one. During much of its life *Dictyostelium* lives as an amoeba in the soil, feeding on bacteria. However, when food becomes scarce, the amoeba sends out a signal, cAMP, a derivative of ATP, which binds to cAMP receptors on other amoeba. The cAMP receptor, a member of the GPCR family, signals to each amoeba, which then moves up the cAMP concentration gradient. The amoeba aggregate, then transform into a fruiting body containing stalk cells and spores. The spores blown from the fruiting body and land some distance away. They lay dormant until conditions are right, and then they become amoeba again. Because of the change from single cells to a multicellular organism, *Dictyostelium* has become a very useful model to investigate the development of multicellularity.

Sponges are among the most ancient multicellular organisms. They consist of two layers, the ectoderm and the endoderm, separated by an extracellular matrix. Sponges have a mouth that contains flagellated cells, which cause small protists to come in contact with it. The protists are then swallowed and digested, and the undigested material is excreted through an anal opening. Sponges do not have a nervous system but contain genes coding for a variety of pre and postsynaptic proteins. Some of these proteins are found in contractile collar cells in the sponge mouth.

Hydra are among the most ancient organisms with true nervous systems. They possess groups of neurons and muscle-like cells that form synapses very similar to neuromuscular junctions. Hydra have a central group of neurons, that is, a ganglion, in the head.

Hydra's genome contains over 20,000 genes, almost as many as the genomes of humans and other mammals. They have a large number of receptors, including over 500 GPCRs. Hydra also secrete over 800 small peptides, which serve as both neurotransmitters and hormones. It is likely that Hydra use neuropeptides as their primary neurotransmitters. In contrast, vertebrates mainly employ small neurotransmitters, such as glutamate, which bind to ionic receptors and transmit signals much faster than do neuropeptides, which signal through slower-acting GPCRs.

At later times in evolution there was a separation between invertebrates and vertebrates. Invertebrates' neurons transmit electrical signals that cause presynaptic endings to release neurotransmitter molecules. Neuronal axons that need to carry electrical signals for long distances accomplish this task by greatly increasing the diameter of their axons.

Vertebrates employed a different strategy. They insulate their long axons with a wrapping made up of many lipid-rich membranes, called myelin. This insulation allows long axons to carry electrical currents for long distances without an increased diameter. This in turn enables vertebrates to have many more neurons in their brains than equivalent invertebrates and allows the creation of the big brains seen in vertebrates, especially in birds and mammals.

During development of the CNS, neurons send out processes whose receptors seek out those with complementary trophic molecules. The ends of these processes contain a specialized structure called the growth cone. The growth cone integrates many signals through their signaling receptors. This allows them to change directions and migrate appropriately through the jungle of developing neurons, processes, and glia. The growth cones contain contractile molecules, actin and myosin, that are able to turn the growth cone in an appropriate direction.

The growth cone uses a variety of second messenger molecules, including increased cytosolic Ca^{++}, cAMP, and cGMP to control the contractile apparatus. New membrane that is made in the cell body is also added at the growth cone through vesicle fusion with the plasma membrane, allowing the neuronal process to elongate.

Once the appropriate contacts are made between processes, synapses are created. The presynaptic ending contains a large number of synaptic vesicles that are filled with a variety of neurotransmitters. A portion of these vesicles lie almost in contact with a specialized region of the presynaptic membrane called the active zone. The active zone contains many Ca^{++} channels.

When a depolarizing electrical charge enters the presynaptic ending, it causes the $Ca++$ channels to open, raising the level of Ca^{++} to a millimolar level. This in turn produces fusion of the synaptic vesicle membranes with the presynaptic plasma membrane, releasing the neurotransmitter(s) contained in the vesicles.

Synaptic vesicle—plasma membrane fusion is mediated by several proteins called SNARES. The SNARES are composed of 3 proteins, VAMP on the synaptic vesicle membrane, and Syntaxin and SNAP 25 on the plasma membrane in the active zones. The three SNARES form a four-member coiled-coil α-helix that anchors each synaptic vesicle to the active zone. When the Ca^{++} concentration rises, a synaptic vesicle protein called Synaptophysin changes its conformation, thereby triggering synaptic vesicle fusion.

Following synaptic vesicle fusion it is necessary to recycle the synaptic vesicle proteins to replenish the population of these vesicles. Otherwise an actively discharging presynapse would rapidly run out of synaptic vesicles and neurotransmitter release would cease. The recycling mechanism involves the endocytosis of the synaptic vesicle protein-containing plasma membrane followed by the formation of clathrin-coated vesicles that rapidly lose their coats to produce empty synaptic vesicles. These vesicles are rapidly refilled with their appropriate neurotransmitter(s) by an acidic ATPase that actively pumps the transmitter into the vesicle.

Each synaptic vesicle also contains a molecule of a small GTPase, Rab 3.

Rab 3 is a member of a family of Rabs that provide a tag enabling each specific Rab-containing vesicle to associate with its appropriate plasma membrane via binding to its specific receptor. The receptor for Rab 3 is called RIM. The binding of Rab 3 to RIM mediates the initial contact between the synaptic vesicle and the presynaptic plasma membrane. After fusion occurs, the Rab 3 hydrolyzes GTP, is released from the plasma membrane, and after binding to another GTP molecule, becomes associated with another synaptic vesicle.

Once the synaptic vesicle contents are released, the transmitter molecules move through the specialized extracellular matrix and a fraction binds to complementary receptors on the postsynaptic membrane. Directly under the membrane in excitatory neurons lies a dense matrix containing a high concentration of the enzyme, calmodulin kinase. When the concentration of Ca^{++} in the cytoplasm rises, produced by the signaling of the activated receptors, this kinase phosphorylates many additional proteins, setting off a series of events leading to an electrical charge moving toward the cell body.

Both pre and postsynaptic cytosols contain numerous scaffolding proteins. These proteins link to other proteins including receptor molecules during synaptic development. Many of these proteins contain PDZ domains that are specifically recognized by other molecules with binding sites complementary to these domains.

The two synaptic components are held together during initial formation and afterward by specific linker molecules. Important ones include the cadherins, which form homophilic ("like with like") connections with identical molecules. Another set of complementary receptors are members of the neurexin and neuroligin ligand, which interact specifically with one another.

Neurons are the most complex group of cells in the body. Each of the 100 billion neurons in the human brain is unique in both structure and function. Neurons contain three compartments—two of which, axons and dendrites, are unique. Axons can be many feet in length and have a network of microtubules running parallel to the plasma membrane from the cell body to the beginning of presynaptic endings. The microtubules are bundled together by the protein tau, which is only found in axons.

To transmit membrane constituents including synaptic vesicles and mitochondria from the cell body to the presynaptic endings, axons have evolved a specialized transport system called rapid axonal transport. This system transports vesicular membranes from the cell body, employing microtubules as tracks. An ATPase called kinesin hydrolyzes ATP, generating the energy to rapidly propel membranes toward the synapse. Rapid axonal transport proceeds at speeds 100 times faster than transport systems in other cells. Axons as well as dendrites contain a rapid retrograde transport system that moves vesicles from the postsynaptic endings to the cell body. It also employs microtubules as the tracks and another ATPase, dynein, as the locomotive. This transport system also moves an order of magnitude faster than does slow transport. Axonal endings are in some cases more than 50 feet away from the cell body. Therefore they have evolved the ability to synthesize their own proteins using mRNAs, ribosomes, and in some cases Golgi stacks, as well as other components.

Dendrites are unique in having microtubules that are polarized in both directions. The microtubules are bundled together by a binding protein unique to dendrites, MAP2. Dendrites also have a rapid transport system employing microtubules and dynein. Dendrites are also able to synthesize proteins at their postsynaptic endings. The mRNAs coding for these proteins are bound to RNA molecules. This prevents them from being employed for protein synthesis until there is a change in the dendrite's environment.

During neuron development the neuron extends several processes. Only one of them becomes the axon. The others become dendrites. A number of receptor–ligand interactions are involved in this determination.

There are three classes of glial cells in the brain: astrocytes, oligodendrocytes, and microglia. Both astrocytes and oligodendrocytes are derived from radial glial cells after neuronal differentiation has ended. Astrocytes have a myriad of functions. They form a network filling the gaps between neurons and ensheath the synapse, taking up excess transmitter molecules to prevent activation of extraneous synapses. Astrocytes synthesize both glutamate and GABA and release them to neurons via vesicular secretion. They also have an immunologic function, recognizing foreign cells and engulfing and degrading them.

Oligodendrocytes' major function is to form the myelin membrane that wraps many layers around the axon's plasma membranes. Myelin provides the insulation that allows the electrical charge to be transmitted rapidly down the axon.

Microglia are the only cell type not produced in the brain. They are born in the bone marrow, migrate into the brain, and remain there throughout life. Microglia are the major immunologic cells of the brain. They ingest and degrade foreign cells including bacteria. Microglia also recognize and degrade inactive synapses and apoptotic cell fragments, using receptors that recognize surface components of these two elements.

The blood−brain barrier (BBB) forms rather early in brain development. The primary components of the BBB are the capillary endothelial cells, astrocytes whose end feet cause the capillary endothelial cells to develop barrier properties, and pericytes, a contractile cell that surrounds the brain surface of the capillary endothelium, and gives it its contractile tone. The BBB forms incredibly tight junctions composed of claudin molecules that form homophilic connections. Not even a H^+ atom can penetrate the barrier, although it is permeable to oxygen and hydrophobic molecules. The major function of the BBB is to allow the homeostasis of the brain to be maintained in the face of major changes in the periphery. It also prevents the invasion of the brain by toxic compounds and harmful bacteria.

The brain requires glucose for energy and synthetic reactions. Therefore the capillary endothelial cells contain glucose transporters that allow glucose to pass through the cells and enter the brain. Also the brain requires Fe^{+++} to form hemoglobin for energy production as well as to produce active enzymes. Therefore capillary endothelial cells contain transferrin receptors that transport transferrin and its iron cargo through these cells and into the brain. Also there are receptors that transport potentially toxic molecules, including amyloid beta, out of the brain.

Several hypotheses are proposed that provide a framework that allows a partial understanding of how the human brain develops from a few cells into a functional structure composed of a hundred billion neurons, a similar number of glia, and many trillions of synapses. The first postulate is that receptor−trophic factor interactions are a key element in establishing the appropriate synaptic connections necessary for the brain's myriad functions. There are also attractive and repulsive receptor−ligand interactions that guide the neuronal processes through the web of developing neurons, axons, and dendrites to reach their appropriate destinations. An important corollary is that there is a quantal number of a given receptor on a particular neuron that requires a similar quanta of a complementary trophic molecule. An example of this corollary is a neuron that contains three quanta of a specific receptor, likely on different processes, will require three quanta of a complementary trophic factor for optimum health.

The second postulate is that the developing brain must destroy neurons that have made inappropriate connections. The first element in this selection process involves both rapid eye movement (REM) and to a lesser extent, slow wave sleep. REM sleep occurs when the brain's neurons are functionally disconnected from those in the periphery. Then the brain's neurons rapidly fire, discharging their neurotransmitters, which also serve as trophic factors. Neurons that have not made their proper connections do not receive their trophic support.

The third postulate is that cells that have made inappropriate connections are removed by apoptosis, also called programmed cell death. Approximately 50% of the mature neurons in the developing brain die. This removal system is only found in birds and mammals. In other vertebrates that contain many fewer neurons, no selection mechanism is required.

The fourth postulate is that receptor–trophic factor interactions are also required during synaptic pruning, which occurs as the brain develops. Inactive synapses are destroyed during this process. However, neurons having other synapses that supply necessary trophic support can survive until other appropriate synapses are formed. In the mature nervous system this process is called synaptic plasticity and is very important in the formation and maintenance of long-term memory.

The final postulate is that many trophic requirements last through life. Connectional diseases of the brain are those in which a small group of cells in one area of the brain die and the pathology spreads synaptically to other connected regions. Neurons that lie very close to the affected neurons, but are not connected, will be spared.

Age-related neurodegenerative diseases including Alzheimer's, Parkinson's, amyotrophic lateral sclerosis, and Huntington's are connectional diseases. It is possible, however, that specific precursor cell transplants may ultimately treat or even cure these dreaded diseases.

The mature cerebral cortex is composed of six laminar layers. Layer VI is near the ventricular surface in the center of the brain while layer I lies nearest the pial surface. The surface of the cortex is flat in mice. In contrast, the human cortex has many indentations in its surface, or sulci, which allow it to contain a many magnitudes greater neuronal population. In the vertical direction, the cortex is divided into a series of columns with interconnected neurons. Large aggregates of columns comprise different cortical regions, for example, the auditory, visual, and motor cortices.

One class of cortical neuron is the glutamate-containing excitatory pyramidal neurons, which extend long myelinated axons to other cortical regions as well as to other areas of the brain. This class of neuron makes up about 70% of the cortical neurons. The excitatory neurons begin life in a region overlying the ventricular surface. Neuronal precursors called neuroepithelial cells form first and divide rapidly. These cells convert to radial glial cells (RGCs), which extend processes from the ventricular to the pial surface. These processes serve as ladder-like structures, along which developing neurons find their way to the appropriate lamina. The first neurons to differentiate form lamina VI. The other laminar neurons move past the early developing lamina VI neurons sequentially to form laminas V, IV III, II, and I.

The other 30% of the cortical neuron population is comprised of GABA-containing inhibitory neurons that form after the excitatory neurons differentiate to form the six lamina. Their site of origin is deeper in the forebrain, in nuclei called the median eminences. Each one of these nuclei produces a distinct group of inhibitory neurons.

After their initial formation, the inhibitory neurons make a long journey toward the cortex. A number of tropic and trophic receptor interactions guide them along the way. They reach the cortex at the same time that axons of the excitatory neurons are exiting the cortex. The inhibitory neurons follow these axons until reaching the cortical lamina, at which time they intersperse throughout the six lamina. The inhibitory neurons each contain a specific neuropeptide-containing vesicle population as well as GABA-containing vesicles. The inhibitory neurons make local synaptic connections with excitatory neurons within the same column as well as with one another.

Many trophic and tropic interactions are necessary for the formation of the precise neuronal distribution within the cortical lamina. One protein that is key to lamina formation is Reeler, a secreted glycoprotein that binds to its receptor, a member of the lipoprotein receptor family. When Reeler is mutated, it causes a staggering gait. Sections of the mutated cortex, when examined, do not have the characteristic laminar appearance. Rather, the neurons do not show any pattern whatsoever. Other mutations of proteins involved in tropic or trophic interactions produce similar cortical disorganization.

As described briefly in a previous paragraph, REM and to a lesser degree, slow wave sleep, are necessary to select the proper circuitry after neuronal differentiation and synaptogenesis are completed. At the beginning of a REM sleep cycle, the neurons of the brain are functionally disconnected from the periphery. The brain neurons begin to fire very rapidly, releasing a large portion of their neurotransmitters. Neurotransmitter release supplies the synaptically connected neuronal population with the trophic factors they require. Of importance here is the finding that the fetus is in REM sleep most of the time, and the major period of brain development occurs during this period. Only birds and mammals have REM sleep. Other vertebrates, which have many fewer neurons, do not require a selection mechanism to produce the proper circuitry. They also do not have REM sleep.

Sleep is also required for additional processes. During sleep, the brain decreases in volume. This allows toxins that have built up during wakefulness to be removed from the brain via channels formed by astrocytes, called glymphatic ducts. Memory consolidation also occurs during sleep. Memory of normal events occurs during slow wave sleep and involves the hippocampus, while memory of emotionally charged experiences occurs during REM sleep and involves the amygdala.

After the circuitry of the brain is assembled and REM sleep has been activated, cells that have not made the correct connections—and therefore have not received their necessary trophic support—are eliminated. The process of cell death, which destroys about 50% of the brain's neurons after maturity, is called apoptosis. During apoptosis, the dying cell becomes dark and its chromosomes are fragmented in a characteristic ladder-like manner. The cell is then fragmented into small vesicles without releasing any potentially toxic compounds into the external environment. These membrane vesicles are then endocytosed by microglia and digested.

The molecular mechanism for apoptosis involves several families of proteins. A group of proteases called caspases destroy the proteins of the cell. Members of the Bcl family of small proteins, if activated, trigger the death process. Molecules called Apaf-1 serve to bind together various proteins involved in apoptosis. Mitochondria become leaky and no longer are able to synthesize ATP. Together all of these factors lead to the programmed destruction of the cell.

Brain-derived neurotrophic factor, a member of the neurotrophin family, plays many key roles in CNS development. It binds to its receptor, Trk B, an integral plasma membrane tyrosine kinase, which is activated after binding. BDNF is involved in the initial patterning of the brain into compartments such as the forebrain, midbrain, and hindbrain.

During early neuron development BDNF plays an important role in determining which of several neuronal processes will become the axon. Later in development it serves as both a tropic and trophic factor. Also BDNF plays a major role in synaptogenesis. After development ceases, it plays an important role in both early and late long-term potentiation, which are necessary for memory formation.

Marijuana is a plant that grows in almost every climate. It has been used for many years as a hallucinogen and for its medicinal properties. Endocannabinoids that mimic the properties of marijuana are found in all regions of the body but are most abundant in the brain. In a mature brain a precursor phospholipid is found on the outer leaflet of the postsynaptic endings. When the postsynaptic ending is activated by neurotransmitter release from the presynaptic ending, the cannabinoids are cleaved from their precursor and bind to one of two GPCRs, CB1 and 2. These receptors are located on the presynaptic membrane. CB1 is the major endocannabinoid receptor in the brain and is its most abundant GPCR.

In the mature brain the endocannabinoids serve to inhibit neurotransmitter release from the presynaptic membrane. They can be thought of as the yang to BDNF's yin. While BDNF binding to the presynaptic membrane causes greater neurotransmitter release, endocannabinoid binding produces inhibition of further release.

During brain development, endocannabinoids play several important roles. They help to specify neuronal identity and to determine which process becomes the axon. They are involved in process growth and also in synaptogenesis.

Besides BDNF and cannabinoids, there are many other proteins that play major roles in brain development. The EphRs, members of the plasma membrane tyrosine kinase superfamily, and their ligands, the Ephrins, were first isolated as proteins that are important for proper retinal ganglion cell axonal guidance into the brain. These proteins mediate repulsive interactions that prevent inappropriate process formation. The EphRs and Ephrins are also involved in early compartment specification, cerebral cortex lamination, axon determination, synaptic specification, and determination of the appropriate number of neurons in the brain.

The Semaphorins and their receptors, the Neuroplexins, are important in early specification of the boundaries of compartments formed during the earliest period of brain development. Their interactions are also repulsive. The Semaphorins and their receptors are also involved in the determination of dendrite formation. They play major roles in lamina specification in the retina and in synapse specification.

The Netrin family and their receptors, the Neuropilins, are attractive. They play important roles in early neuronal determination and synaptogenesis.

Proteoglycans and glycoproteins are secreted molecules that are important in the formation and maintenance of the extracellular matrix (ECM). The ECM is crucial during brain development for neuronal migration, axonal and dendritic guidance, and synaptogenesis.

Once a given neuronal process reaches its approximate destination, it must form a synapse with the correct neuron. There are two types of receptors that play major roles in terminal specification.

The protocadherins are members of the cadherin superfamily and contain several CAD domains in their extracellular regions. Protocadherin genes are clustered into three groups. Combinations of protocadherin monomers from each group form tetramers on the presynaptic surface, which make homophilic interactions with an identical tetramer on another neuron's postsynaptic surface. Because there are more than 10^{12} possible tetrameric combinations, it is very likely that they play major roles in terminal specification.

The other major protein family that plays a major role in terminal specification in the olfactory system are the odorant receptors (ORs)—the largest group of receptors in the mammalian genome. Each neuroepithelial cell (NEC), which is a specialized neuron in the nostril, contains only one OR on its cell body and axon. There are enough different ORs on the NECs to detect more than 100,000 different odors.

When the axon of a particular NEC reaches the olfactory bulb in the brain, it forms a synapse with a specific olfactory bulb neuron. Several different NEC axons form synapses on a particular neuron, which allows for integration of several odors in the bulb. Many of these synapses are formed and then the information travels to the brain where further integration occurs. Although ORs are important in synapse specification, other proteins play a role, including the protocadherins.

There are two sex hormones: estrogen, made by the ovaries, and testosterone, produced by the testes. These hormones are secreted into the bloodstream. When they reach the brain, because they are very hydrophobic, they enter neurons that contain either estrogen or testosterone receptors, which are scattered throughout the brain. Once the hormone binds to its specific nuclear receptor, the receptor dimerizes and activates genes that code for proteins conferring either masculine or feminine traits.

Glucocorticoid receptors are located in variety of neurons. They are the only steroid hormone receptor located in the cytoplasm. When bound to a glucocorticoid, the receptor enters the nucleus, dimerizes, and leads to the production of proteins that act on neurons in the hypothalamus, to minimize responses to stressful events.

Another important steroid is vitamin A. It is essential for many processes that occur during brain development. It can also be converted into retinol, which binds to amino acids in the channel formed by the opsin protein, similar in structure to Bacteriorhodopsin, to form rhodopsin, the visual pigment. When a photon of light impinges on the *trans* form of retinol, it is converted to the *cis* form, activating the photoreceptor cells. These cells carry visual information through different neurons. Ultimately the axons of retinal ganglion cells convey visual information to the brain. Children who do not consume sufficient vitamin A become blind. This is a major health problem in developing countries.

The hypothalamus synthesizes two groups of hormones that are responsible for a variety of behaviors. Two of these molecules are the very similar 9−amino acid containing peptides oxytocin and vasopressin. When secreted into the blood, they bind to their respective receptors. These two hormones play important roles during brain development, because they bind to GPCRs on many brain neurons, thus altering a variety of behaviors, including social interactions.

The hypothalamus secretes a variety of very small peptides called releasing factors, which empty into veins directly above pituitary neurons. Among these releasing factors is growth hormone−releasing factor, which activates its GPCR to secrete growth hormone, in turn stimulating growth of the body and the brain. Another factor is thyrotropin-stimulating releasing factor. After secretion, thyrotropin binds to GPCRs on thyroid cells, inducing the release of thyroxine, which is the master metabolic hormone in the body. An important releasing factor is corticotropin-releasing factor, which after binding to its GPCR, triggers secretion of adrenal corticotropin, a major stress-mediating hormone. There are also two inhibitory releasing factors secreted by the hypothalamus: growth hormone inhibitory releasing factor and prolactin inhibitory factor. All of these releasing factors are produced by neurons in the brain and modify both development and behavior.

The enteric nervous system (ENS) develops from a small group of neuronal precursors that migrate as a chain of cells, ultimately surrounding the lining of the gut and innervating it. The migration of the enteric nerve cells depends on a variety of trophic and tropic factor−receptor interactions. Among the most important of these is the binding of glial cell−derived growth factor to its receptor, Ret. Individuals with mutations in either of these proteins develop Hirschsprung disease, leading to an incomplete innervation of the intestinal tract and difficulty eating.

At birth, the entire intestine is colonized by millions of microorganisms and viruses. This colonization forms the microbiota. The genomes of these protists constitute the microbiome. The microbiota supplies important nutrients, especially small fatty acids, to both the developing and mature brain. It also produces a variety of neurotransmitters whose number exceeds that of the brain. These neurotransmitters influence both the gut and indirectly, the brain.

Fecal transplants are now employed to treat *Clostridium difficile* infections in individuals whose intestinal flora has been destroyed by antibiotics. These transplants have also been used to treat a variety of other conditions.

Viruses, while of value to the microbiota, can produce devastating diseases in the CNS. Among these viruses are poliovirus, which infects motor neurons and causes paralysis; herpes simplex virus, which destroys large numbers of neurons in the hippocampus; Zika virus, which destroys radial glia in the developing brain, producing small brains, intellectual disabilities, and death in some cases; and rabies virus, which damages many neurons in the brain and eventually produces death.

Synaptic pruning is the mechanism by which the brain eliminates inactive or damaged synapses. This process is necessary because synapses become inactive or damaged, and if not removed, can interfere with the brain's circuitry.

One may ask, why is it that a neuron, deprived of its normal trophic support by loss of a synapse that provides trophic support, can survive? This question may be explained by the corollary to the first hypothesis presented in Chapter 7, a testable theory to explain how the wiring of the brain occurs; namely, that if a neuron has three quanta of a particular receptor, it requires three quanta of the complementary trophic factor to survive. If a neuron loses one quantum of a trophic factor, presumably by loss of the synapse that provides it, the neuron can send out another process together with upregulating the receptor for the trophic factor it requires. The upregulated receptors allow the process to follow a gradient of the required trophic factor, until the process contacts another neuron producing the required trophic factor, making a new synapse and rescuing the tropically disconnected neuron. A similar process occurring in the mature brain is called synaptic plasticity.

There is now evidence concerning the mechanism of synaptic pruning. Two proteins, C1 and C3, which are part of the immunologic surveillance system, tag the surface of inactive or damaged synapses. Microglial surface receptors recognize these proteins and the microglial cell endocytoses and degrade the tagged synapses.

Abused drugs are perhaps the largest health problem our society faces. The opioid epidemic now kills more people than automobile accidents. Drugs of abuse are separated into legal and illegal drugs. It is an arbitrary distinction that prevents many people from obtaining needed support to cope with addiction, while filling our prisons with many people who are convicted of possession of a small quantity of an illegal drug.

All abused drugs work, at least in part, through the same mechanism. They all stimulate the pleasure center in an area of the brain called the nucleus accumbens, which lies in the basal forebrain. Dopaminergic neurons from the ventral segmental area interact with excitatory neurons from the cerebral cortex. These neurons stimulate dopamine release from the nucleus accumbens' synapses, evoking a pleasurable sensation. With time, however, the accumbens' dopamine receptors are upregulated, requiring more of the drug to produce the same effect. This phenomenon is called tolerance. Tolerance leads to a craving for the drug, producing dependence. Also withdrawal from the drug induces a whole host of unpleasant effects. Therefore the individual becomes addicted, constantly seeking the drug at whatever cost to avoid the pain of drug withdrawal.

Among the illegal drugs of abuse are cocaine and amphetamines, which cause the release of excess dopamine from synapses in the nucleus accumbens. Cocaine was thought to damage the fetus when abused by the mother. It has recently been shown, however, that the damage to the fetal brain is minor and is reduced with time.

Opioids, whose use has become an epidemic, work by binding to a family of receptors, all GPCRs, in many regions of the brain. The μ receptors are on inhibitory neurons that synapse with dopaminergic neurons in the nucleus accumbens. When an opioid binds to this receptor, it blocks GABA release, in turn producing increased dopamine release from the neurons in the pleasure center. While there is no apparent damage to the fetal brain, the babies of mothers abusing opioids are also opioid dependent and must remain hospitalized while undergoing painful withdrawal. There is evidence that these babies show some intellectual impairment and behavioral abnormalities as they develop.

Two legal drugs cause great damage to health. Alcohol binds to excitatory and inhibitory receptors on neurons synapsing with those in the nucleus accumbens. This invokes a transient pleasurable feeling. Soon, however, alcohol's binding to inhibitory synapses in other regions of the brain produces depression and drowsiness, leading to many incidents of date rape and automobile accidents. Long-term abuse also damages the brain and other organs. Babies born to mothers who abuse alcohol during pregnancy suffer from fetal alcohol syndrome. This condition leads to both impaired intellectual ability and behavioral abnormalities.

Tobacco is a widely used drug of potential abuse. Nicotine, the most active compound in tobacco, binds to nicotinic receptors in many regions of the brain including the nucleus accumbens, leading to a pleasurable relaxed sensation. Moderate to heavy use leads to many damaging effects and is a leading cause of premature death. Smoking during pregnancy produces smaller babies with development, intellectual, and behavioral deficits.

All of the drugs described here, with the possible exception of tobacco, have powerful indirect effects on child development. The drug-abusing parent devotes much of his or her time to obtaining and consuming the abused drug(s), causing economic harm to the child as well as either physical abuse or neglect.

Many developmental brain diseases have a genetic origin and involve receptors and/or trophic factors. Fragile X disease is caused by amplification of the codon CGG in the gene for the FMR1 protein. This protein's function in neurons is to bind specifically to a group of mRNAs in the nucleus. This binding, which is specific and selective, can be thought of as comparable to receptor–ligand binding. Once this interaction has occurred, FMR1 escorts these mRNAs from the nucleus to the cytoplasm and then transports them to the postsynaptic endings. An excess number of contiguous CGGs produces a malfunction of the FMR1 protein. This malfunction in turn produces intellectual and social deficits.

Down syndrome is caused by the triplication of chromosome 21 during egg, or less commonly, sperm development. The triplication causes overproduction of several receptors leading to intellectual and behavioral impairment. Among the proteins that are overproduced is the β-amyloid precursor protein (APP). APP binds to a variety of protein-containing vesicles and via its connection to either kinesin or dynein, transports these molecules to either the presynaptic endings or the cell body. One of the proteins transported from synapse to cell body is NGF. NGF serves as a trophic factor for neurons in the nucleus basalis in the forebrain. These neurons produce acetylcholine and synapse with many other neurons in the cerebral cortex and other areas. Excess APP interferes with NGF transport and destroys many of these cholinergic neurons, which play an important function in memory. As individuals with Down syndrome are living well into their forties and fifties, the excess APP causes them to develop Alzheimer's disease. This is thought to be caused by the overproduction of the amyloid beta peptide (Aβ), a breakdown product of APP.

Bipolar disease and schizophrenia are two "psychiatric" diseases that usually begin during the late stages of brain development. Both diseases occur in higher frequencies in families than in the general population. This suggests that they have genetic causes; although the genetics of both diseases is complex, indicating the interaction of many gene variants.

Bipolar disease is treated with lithium or valproic acid, both of which activate neurotrophic and neuroprotective signaling pathways. Schizophrenia is treated with many antipsychotic drugs that bind to dopaminergic, serotonergic, and Glu receptors. These receptors signal to produce neuroprotective proteins. There is also evidence that schizophrenia may be caused by excess synaptic pruning during the latter stages of brain development.

Autism spectrum disorder (ASD) produces social and other behavioral abnormalities. Two neurohormones, oxytocin and vasopressin, which are very important in the development of appropriate social interactions, are decreased in ASD patients.

In 1981 Stanley Appel coined the term "connectional diseases" to describe a group of age-related neurodegenerative diseases. All of these diseases begin in a specific region of the brain and progress sequentially to other synaptically connected regions. Loss of trophic support of one neuron caused by the death of a connected partner produces this effect. Neurons adjacent to affected neurons, if not synaptically connected, are spared. Among these devastating diseases are Alzheimer's disease (AD), Parkinson's disease (PD), Huntington's disease (HD), and amyotrophic lateral sclerosis (ALS). Two other conditions meet some criteria for being categorized as connectional diseases: chronic traumatic encephalopathy (CTE) and prion diseases.

AD was first described in a 41-year-old woman with severe dementia, AD is characterized by three lesions. On gross inspection the brain is shrunken, with enlarged ventricles. The other lesions are visualized on sectioned brain using histochemical staining. These pathologic lesions are Aβ-containing extracellular plaques and intracellular neurofibrillary tangles (NFTs) composed mainly of the tau protein. Individuals with AD first show memory loss, followed by inability to carry out activities of daily life, which progresses until death.

There is strong evidence that Aβ initiates AD. Aβ-containing plaques are seen well before any clinical symptoms are evident. Several mutations either within the Aβ sequence or within the sequence of an enzyme, secretase, also produce AD. There is one mutation in the Aβ sequence that makes individuals with the mutation less likely to get AD. Individuals having the E4 variant of the Apolipoprotein E (ApoE) have a much greater risk of getting AD than do those with the E2 or 3 variant. ApoE binds to Aβ molecules in the brain and removes them. ApoE4 does not remove Aβ as well as do the other variants.

The intracellular NFTs are predominantly composed of aggregates of hyper-phosphorylated tau, The NFTs are the likely pathological initiators of AD. Tau aggregates disrupt many cellular processes, leading to death of the affected neurons. The spread of cells with NFTs follows exactly the clinical progression of the disease.

PD produces a variety of motor symptoms including rigidity, slowness of gait, and at later stages, inability to swallow followed by death. PD is histochemically characterized by synuclein-containing nuclear deposits in a small group of dark pigmented neurons in the substantia nigra. These neurons secrete dopamine, and when they begin to die, early PD symptoms appear. PD is the first age-related neurodegenerative disease shown to be caused by decreased release of a specific neurotransmitter: in this case, dopamine. It was also the first of these diseases for which a treatment was developed. It was shown that a precursor to dopamine, L-dopa, can readily cross the BBB, enter the nigral dopaminergic cells, and replenish their vesicular dopamine content. The treatment alleviates the motor symptoms for approximately five years. After this period, sufficient dopaminergic cells have died to cause disease progression.

There are a variety of genetic forms of PD. Mutations in synuclein produce the disease. Also mutations in proteins involved in damaged mitochondria removal or in endocytosis can produce the disease.

ALS and HD are connectional diseases leading to death of motor neurons and striatal neurons respectively. ALS destroys motor neurons in a variety of ways.

Mutations in several different proteins produce genetic forms of the disease. HD is caused by large increase in the number of CAG repeats in the gene coding for Huntingtin. This protein has a myriad of functions and when an excess number of repeats is reached, the protein is unable to carry out its functions.

CTE is caused by several mild concussions leading many years later to behavioral changes including apathy, depression, aggression, and later, dementia and death. CTE may be a connectional disease. Initially there are NFTs in neurons surrounding blood vessels near the pial surface. With time these lesions spread to other regions of the brain that may be synaptically connected. The symptoms of the disease progress in a gradual manner as well.

Prion diseases are caused by mutant proteins that can convert normal proteins into abnormal ones. The prion diseases, in contrast to the other diseases already mentioned, have been shown to be infectious and can be transmitted from one species to another as is the case with mad cow disease. In this case the prion that infects cattle can also infect humans who eat meat derived from cattle suffering from the disease.

While at this time neurodegenerative diseases of aging are incurable, there is hope that this situation may soon change. The pcd mouse has Purkinje cells that mature and then begin to die. It has been demonstrated that, if a cerebellar transplant at the precise stage of development is placed over the pcd cerebellum, the Purkinje cell precursors can migrate deep into the cerebellar cortex and replace the dying Purkinje cells both anatomically and functionally.

At the same time, surgeons were implanting dopaminergic precursors into the region of the substantia nigra of patients with late-stage PD. These transplants did not work very well. This was likely due in part to immunologic differences between host and transplanted cells. Also the precise developmental stage in which the transplant would prove efficacious was not known.

In animal models, regions of the brain in which interneurons had been destroyed were given transplants containing GABAergic precursor cells. These cells migrated into the damaged brain region, made synapses, and produced anatomical and physiologic improvement. The precursors migrate into the brain and make synapses following their own developmental schedule, with no effects from the recipient brain.

Recently a major breakthrough has occurred. It has been demonstrated that either white blood cells or skin fibroblasts taken from an adult can dedifferentiate to totipotent cells and be transformed into neuronal precursors. These cells can then be differentiated to form specific classes of neurons. Already this technique has been employed to examine differences between cortical neurons derived from patients with AD and other neurodegenerative diseases with those from normal individuals. Drugs that hopefully normalize diseased cells can also be screened.

In the future it should be possible to transplant normal neuronal precursors into the brains of patients with neurodegenerative diseases to arrest disease progression or even "cure" the disease. An advantage of this technique is that the patients' own cells will be transplanted, avoiding immunologic differences with the host brain. One disadvantage is that the transplanted cells will contain the mutated gene(s). In the case of those patients with age-related connectional diseases, this should not be a problem because the disease phenotype is only expressed in later life.

20.2 CONCLUSION

This book has taken us on a journey from the most primitive organisms to the most complex machine ever evolved, the human brain. Initially receptor—ligand interactions were used as signaling systems to alert the organism to change their environments to locate food or escape prey.

After multicellular organisms evolved, receptor—ligand interactions took on the additional function of aiding the development of the proper circuitry for primitive nervous systems. With the evolution of vertebrates, these interactions play key roles in brain development and function. The human brain represents the most complex and dynamic computer in the universe. It can do amazing things including writing poetry, composing music, designing large buildings, etc. Unfortunately it can also build enormous weapons with the capability of destroying the Earth. Also, humans have polluted both the air and sea, thereby risking environmental destruction. Perhaps in creating the human brain, evolution has taken a wrong turn. Hopefully, humans can learn to adapt in healthy ways to caring for their environment and one another.

Index

Note: Page numbers followed by "*f*" refer to figures.

Printed in the United States
By Bookmasters